LOCAL ORGANIC

Local Organic

FOOD RHETORICS AND COMMUNITY WRITING FOR IMPACT

Veronica House

Utah State University Press
LOGAN

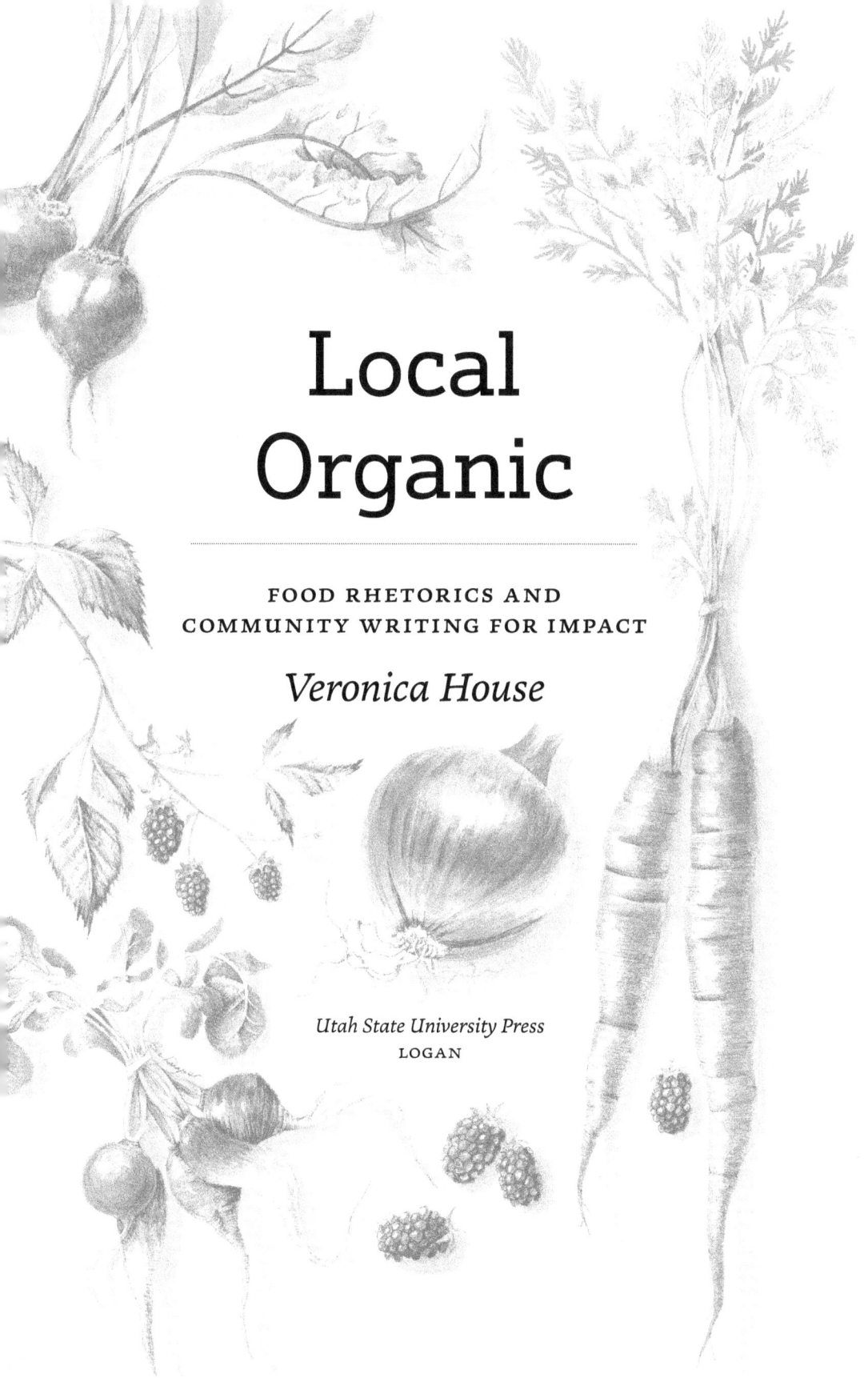

© 2025 by University Press of Colorado

Published by Utah State University Press

An imprint of University Press of Colorado
1580 North Logan Street, Suite 660
PMB 39883
Denver, Colorado 80203-1942

All rights reserved

 The University Press of Colorado is a proud member of Association of University Presses.

The University Press of Colorado is a cooperative publishing enterprise supported, in part, by Adams State University, Colorado School of Mines, Colorado State University, Fort Lewis College, Metropolitan State University of Denver, University of Alaska Fairbanks, University of Colorado, University of Denver, University of Northern Colorado, University of Wyoming, Utah State University, and Western Colorado University.

ISBN: 978-1-64642-718-5 (hardcover)
ISBN: 978-1-64642-719-2 (paperback)
ISBN: 978-1-64642-720-8 (ebook)
https://doi.org/10.7330/9781646427208

Library of Congress Cataloging-in-Publication Data

Names: House, Veronica, 1974– author.
Title: Local organic : food rhetorics and community writing for impact / Veronica House.
Description: Logan : Utah State University Press, [2025] | Includes bibliographical references and index.
Identifiers: LCCN 2025001822 (print) | LCCN 2025001823 (ebook) | ISBN 9781646427185 (hardcover) | ISBN 9781646427192 (paperback) | ISBN 9781646427208 (ebook)
Subjects: LCSH: Communication in ecology. | Local foods—Social aspects. | Natural foods—Social aspects.
Classification: LCC QH541.18 .H68 2025 (print) | LCC QH541.18 (ebook) | DDC 641.3/02014—dc23/eng/20250221
LC record available at https://lccn.loc.gov/2025001822
LC ebook record available at https://lccn.loc.gov/2025001823

Cover illustration by Phoebe Draper

Contents

Acknowledgments *vii*

Land Acknowledgment *xi*

Introduction: Ecological Community Writing and the Power of Definitions *3*

On Lineage and Purpose: A Place to Begin *31*

1.
Food Rhetorics and a Crisis of Definitions *39*

2.
What's "Local" About Local Food? Rehinging Words to Meaning *75*

3.
Crisis and Abundance Rhetorics in Moving People Toward Collaborative Engagement *107*

4.
Writing Abundant Ecologies: From Theory to Collaborative Local Research *139*

5.
Ecological Community Writing in an Abundant Foodshed
with Constance Gordon *170*

6.
Abundant Partnership: A Local, Organic Approach
to Rhetoric and Writing Studies
with Kerry Francis, Tim Francis, and Kelly Zepelin 218

Works Cited 249
Index 271
About the Author and Contributors 287

Acknowledgments

As I have written this book over the last seven years, I have watched out the window from my home desk, learning the area residents. Many families of deer have moved slowly through the grasses behind our house and lazed behind the fence, unfazed by our dogs' angry barking. Countless birds have built nests and flittered in the spruce tree in the yard. The coyotes have occasionally made a home out back and loosed eerie howls when they catch prey. The red-tailed hawk has sat patiently every day in the tree beyond the fence. Night after night, red foxes have stalked the land to the east of our house. One spring, a moose wandered down from the mountains and lived in the flood plain behind us for a week or two. And two black bears camped out in the back for a time. Thank you for filling my life with beauty and wonder.

I owe thanks to so many colleagues who have supported and helped shape my ideas and dreams through the years. You have helped me manifest what we've been able to accomplish together, and your scholarship and friendships have been a solace and a warm fire to keep me going. For those of you who have impacted this book project in particular, I am so grateful. Paula Mathieu, thank you for reading an early version of some of these chapters and for reminding me to believe in *abbondanza!* Rachael Shah,

Stephanie Wade, Iris Ruiz, thank you for reading an early draft of a chapter or the full manuscript and offering careful guidance and encouragement as ideas took shape. I appreciate you for it! Paul Feigenbaum, we talked about an early idea for this book that encouraged me to keep writing. Getting to know you and watch your thoughtfulness as a reader and an editor over the last several years has been a wonderful experience. Seth Myers, your friendship sustained me through my years at CU, and you were my partner and accomplice in so many exciting ideas. Dreaming big with you at Fate was so much fun and a highlight of my career. I'll never forget our excitement as we began to imagine the first Conference on Community Writing and then the second! Thank you! Tobi Jacobi, I wrote a lot of this book with you in libraries and coffee shops. You are such a dear friend, and you motivated me again and again to keep writing. I appreciate your encouragement, as well as the jams, breads, and other treats you've gifted through the years. Carmen Kynard, I hold close your words to *do the work*. I try my best to live into your reminder every day. Thank you for being such a generous and thoughtful colleague. Beth Godbee, thank you! Thank you for coaching me and challenging me to ask the difficult questions. Calley Marotta, thank you for your friendship and for your sage advice as you read this book in its final stages. This book is stronger because of your recommendations and insights.

Kerry Francis, how lucky we were to find you as a preschool teacher who opened up a beautiful world of wonder for our girls. Tim Francis, you and Kerry have changed our lives. You've offered meadows and trails, songs and joy, festivals and light. And the food, all that delicious food! I'm grateful for you both.

To the former members of The Shed—those years working with you were a joy. Thank you especially, Tim Plass, for inviting me into the group. And a huge thank you to those on the board I worked most closely with and who contributed your expertise to the experiences shared in this book—Carl Castillo, Ann Cooper, Micah Parkin, and Brian Coppom.

Thank you, Ashley Dancer, Constance Gordon, Peter Newton, and Phaedra Pezzullo for partnering with me on Dig In! I loved collaborating with you. I owe a big thank you to the Office of Outreach and Engagement at CU Boulder for the Dig In! grant that allowed for much of the work described in this book to take place. Thank you to the former students from Boulder High School who were so kind to allow me to feature their art in this book.

Kelly Zepelin, I hope someday that you will take me foraging with you! Thank you for all you taught me and for writing with me. Phoebe Draper, thank you for the gift of your beautiful artwork for the book's cover. I'm so lucky to have met you at Dharma's Garden.

Sheila Carter-Tod, thank you for your support of my work and for embodying daily what compassionate and generous leadership looks like. Isabel Baca, Stephanie Briggs, Fatuma Emmad, Eli Goldblatt, Megan Hartline, Ada Hubrig, Megan Kelly, Lindsey Loberg, Heather Martin, Madeline Nall, Kasey Neiss, Carmen Pacheco-Borden, Elaine Richardson, Ross Rodgers, Lauren Rosenberg, Donnie Sackey, Dan Singer, and so many more—each of you uniquely impacted me and taught me at critical moments during the process of writing.

Rachael Levay, Nate Bauer, Skylar Cooper, Dan Pratt, and the whole team at Utah State University Press and University Press of Colorado, thank you for believing in this project and for your help in bringing it into the world!

I couldn't have done this work without my undergraduate writing students, who year after year came up with such wonderful project ideas. Thank you for being *the best* students!

Jessie, Erik, Zoe, Mom, Dad, you bring so much laughter, joy, and support to my life. Jessie, there are no words for how important you are. I love you always. Kevin, where do I begin? You read every page of this book multiple times (even when you didn't want to!) and patiently and generously offered support. You took care of the girls and the house and the dogs when I was holed off writing. You believe in me and my ideas and have helped to shape so many of them. Thank you! Cassidy and Allegra, you are my everything. I love you up the willow tree and down the willow tree, infinity times infinity.

Thank you all for making my life so full.

Land Acknowledgment

Because this book is about place, including the legacies of colonialism and racism within our food systems, and my work as a community-engaged writing professor living in Boulder, Colorado, I want to open first with the City of Boulder Staff Land Acknowledgment for the land that provides me a home and nourishment, as well as a place from which I write this book. I value this land acknowledgment because it was written collaboratively with American Indian Tribal Nations and the Boulder community and because it offers both acknowledgment and a commitment to action. I am grateful to the City of Boulder for permitting me to reprint their current acknowledgment, understanding that "the acknowledgment may change over time as we continue to build and strengthen relationships with Tribal Nations and Indigenous communities" (Yates).

***

The City of Boulder acknowledges the city is on the ancestral homelands and unceded territory of Indigenous Peoples who have traversed, lived in, and stewarded lands in the Boulder Valley since time immemorial. Those Indigenous Nations include the: Di De'i (Apache), Hinono'eiteen (Arapaho),

Tsistsistas (Cheyenne), Nʉmʉnʉʉ (Comanche), Kiowa, Čariks i Čariks (Pawnee), Sosonih (Shoshone), Oc'eti S'akowin (Sioux) and Núuchiu (Ute).[1]

We honor and respect the people of these Nations and their ancestors. We also recognize that Indigenous knowledge, oral histories, and languages handed down through generations have shaped profound cultural and spiritual connections with Boulder-area lands and ecosystems—connections that are sustained and celebrated to this day.

The City of Boulder recognizes that those now living on these ancestral lands have a responsibility to acknowledge and address the past. The city refutes past justifications for the colonization of Indigenous lands and acknowledges a legacy of oppression that has caused intergenerational trauma to Indigenous Peoples and families that includes:

- For more than 10,000 years, generations of Indigenous Peoples have lived and thrived on ancestral homelands that Euro-Americans colonized as Boulder.
- Indigenous Peoples in Boulder have, as in all parts of the Americas, endured centuries of cruelty, exploitation and genocide.
- The westward expansion of Euro-American population and culture in the nineteenth century caused extensive hunger and diseases that devastated Indigenous Peoples' way of life.
- In October 1858, Hinono'ei neecee ("Arapaho Chief") Nowoo3 ("Niwot," "Lefthand") told a party of gold-seekers camped in what is now known as Boulder that they could not remain on Indigenous land as defined by the 1851 Treaty of Fort Laramie.
- After gold was found west of Boulder in January 1859, many of those same gold-seekers helped found the Boulder Town Company on Feb. 10, 1859, in violation of the 1851 Treaty of Fort Laramie.
- By the summer of 1859, thousands of gold seekers were in the Boulder area, and many squatted on Indigenous lands, continuing the dramatic expansion of Euro-American occupation of Indigenous lands that soon exiled Indigenous peoples from the Boulder area.
- In August 1864, more than 100 Boulder County residents mobilized into Company D of the Third Colorado Cavalry at Fort Chambers along Boulder Creek east of what is now known as Boulder.

---

1 Names are based on interviews with Tribal Nation Representatives from June 2021 through February 2022. Please note: The appropriate Kiowa Indigenous name has not yet been obtained from Tribal Representatives.

- Company D—which included 46 Boulder men and prominent Boulder County residents—later participated in the barbaric massacre of peaceful Tsistsistas and Hinono'eino' at Sand Creek on Nov. 29, 1864. Among those killed in the massacre were women, children, elders, and chiefs, including Now003 and Tsistsistas Chief White Antelope. Despite having participated in horrific atrocities, members of Company D received a heroes' welcome upon their return home.
- The city has benefited and continues to benefit directly from the colonization of Indigenous lands and from removal policies that violated human rights, broke government treaties, and forced Indigenous Peoples from their homelands.

We must not only acknowledge our past but work to build a more just future. We are committed to taking action beyond these words. We pledge to use this land acknowledgment to help inspire education and reflection and initiate meaningful action to support Indigenous community members and our federally recognized American Indian Tribal Nation partners.

We intend to use this acknowledgment when the City of Boulder develops work plans that guide day-to-day work, begins new projects, starts long-term community plans, and recruits and hires staff.

Let this formal acknowledgment—which honors and builds on the city's Indigenous Peoples Day Resolution (1190)—stand as a critical step in our work to unify Boulder communities, combat prejudice and eliminate discrimination against Indigenous Peoples.

* * *

In addition to the City of Boulder Staff Land Acknowledgment, which provides an important historical overview and proposed ongoing commitment, I would like to acknowledge further that I currently work at the University of Denver (DU), which resides on unceded lands that were stewarded by the Ute, Cheyenne, and Arapaho peoples. The founder of DU, John Evans, holds a large responsibility for the Sand Creek Massacre. On August 11, 1864, Evans, then superintendent of Indian Affairs in the Colorado Territory, issued a proclamation that authorized white settler colonists of Colorado "to kill and destroy, as enemies of the country . . . all hostile Indians." This proclamation resulted in the Sand Creek Massacre where at least 230 Native Americans were killed, most of whom were children, women, and the elderly. The proclamation was only rescinded in 2021 by Colorado's governor, Jared Polis. Every time I drive to campus, I turn into the university on Evans Avenue, a shameful reminder of the university's founder.

Often led by Native and Indigenous student groups and faculty on campus, the University of Denver community members continue to expose and offer recommendations for how the university can better own up to its history and honor the Native and Indigenous initiatives and demands for the university. I acknowledge and honor the Ute, Cheyenne, and Arapaho Tribal Nations and all the original Indigenous peoples who stewarded these lands. I commit to

- ongoing, lifelong learning about the history of the land and its inhabitants prior to and since white settler colonization;
- teaching in my food studies writing courses about past and current work by Native peoples to maintain and reintroduce Native foodways;
- supporting BIPOC-led farms and organizations doing critical work in Front Range communities such as Frontline Farming in Denver, Harvest of All First Nations in Boulder, and First Nations Development Institute in Longmont.

As I explore throughout this book and explicitly in the "On Lineage and Purpose" section after the "Introduction," this place—this land and its history—has called forth in me a personal and professional commitment. This book is a disciplinary offering of my attempt to dive into what writing in this place means while working toward accountability and repair.

LOCAL ORGANIC

INTRODUCTION

Ecological Community Writing and the Power of Definitions

More than a decade ago, my undergraduate writing students and I asked what we thought was a simple question in our course on food rhetorics: "What is local food?" Answers to this seemingly straightforward question about how local food is defined in our city of Boulder, Colorado, surprised us. Responses across a variety of community members from farmers to restaurant owners to consumers illustrated widely differing ideas, ethical considerations, and visions for a definition of local food. The competing or conflicting definitions sometimes led to consumer confusion as well as difficulty in creating policy and support for local farming. This revelation about definitional inconsistencies and tensions led to questions that launched years of courses, projects, events, scholarship, and food literacy work I have coordinated with students, faculty, and project partners across Boulder County, Colorado.

Local Organic is a book at once about collaboratively building community-derived definitions for resilient local food systems and about building coalitions across academic disciplines and communities through writing. In it, I explore how higher-education faculty and students can work to ethically partner with local communities, and how Rhetoric and Writing faculty and students specifically can support this work through community-based writing, courses, research projects, food literacy events, and coalition building—as parts of a methodology for engagement that I call *ecological community writing*. While Rhetoric and Writing Studies has a decades-long tradition of community-engaged scholarship, in *Local Organic* I offer a relational methodology for classroom and community work to drive curricular design and community-wide writing about social and environmental issues.

Ecological writing studies (M. Cooper, "The Ecology of Writing"; Weisser and Dobrin; Edbauer; Gries, *Still Life*) analyzes the complex, dynamic systems within which writing functions and the circulation and remix of texts within those systems. An *ecological community writing* methodology adds to this previous work by offering a method for how to not only study an ecology but use writing to build connections that strengthen or generate a coalitional ecology of work. The "writing," capaciously defined as not only alphabetic but multimodal and embodied, connects writing programs, courses, and scholarly work to community conversations and social issues. This is done through multiple, distributed writing projects co-developed with community members, writing students, and faculty members across disciplines, with an eye toward social impact and emotional connection. This method, which I call *distributed definition building*, moves people toward engagement in local issues as they generate writing that helps to build understanding and awareness on personal and community-wide levels. Throughout the book, I study and theorize the ecology of work and beliefs around local food in Boulder County from different angles (language, affect, people and organizations, and other-than-humans) to show how each project, course, and partnership makes connections that help to build that ecology.

The term "local food" is not alone in generating confusion. Many contested terms and issues share this definitional problem. For this reason, Rhetoric and Writing teachers and scholars (from here on out mostly referred to as "practitioners" for brevity's sake and to acknowledge our multiple and intertwined roles as community members, activists, organizers, and any of the many ways we may define ourselves) are well suited to intervene in community discourses around definitional conversations and conflicts. While this book deliberately does not offer a singular answer for what "local food" is or a way to easily create uniform definitions about contested terms, *ecological community writing* that uses *distributed definition building* helps community members create personally and culturally meaningful definitions for contested or complicated terms based on local needs and considerations.

Who Holds the Power in Definition Building?

In the chapters that follow, I describe more than a decade of work with my undergraduate writing students and project partners in Boulder County to build communication, food literacy projects, and events about local food and farming. To avoid often confusing top-down definitions of "local food" created only by those in power, we moved away from a more traditional rhetorical concept of a "definition argument," in which the objective is to argue for a single and "best" definition for a term. Instead, I developed *distributed definition building* as a method to foster community literacies around contested terms and social issues. In our case, we encouraged community members to reflect critically on what the term "local food" means to them and to move into action for just, culturally relevant, and accessible local food. In this way, this method of *distributed definition building* offers a richness and complexity of experience and connection, an intervention that is deeply rhetorical and that can be used to study the power of language to create reality in the context of power dynamics.[1]

For our purposes in studying what "local food" means to people and organizations across Boulder County, the hundreds of definitions my students and I gathered suggested that local food offers a deep experience of community that cannot be matched by an industrial (i.e., large-scale, resource-intensive) food experience. It can offer an antidote to structures of injustice built into the industrial food system, which, as this book will explain, result in vast amounts of food waste, food-related disease, labor exploitation and modern-day slavery, and disconnection from one another, the earth, and animals. *Distributed definition building* draws out nuances and variants that can be lost in an imposed, top-down definition.

As I will discuss in chapter 1, the industrial food system's attempts to claim definitional control over language are connected to exploitation on

---

1 Rhetoric and Writing Studies offers studies of naming (e.g., Vanguri) and of expansion of definitions of, for example, what non-western traditions must be included in the field's understanding of rhetoric (e.g., Ruiz; Hass; García and Baca). I am not aware of any study, however, that uses a distributed method of definition building.

many fronts. Indeed, the industrial food system is grounded in the desire to "claim" wealth, land, resources, and power (as in the early "land claims" required of homesteaders and claims to the bodies and labor of African people who were enslaved to farm). It is grounded in hoarding of resources and in uniformity. It is grounded in monoculture and confined animal production that are also rooted in settler colonial claiming of land and living creatures as property to be used for profit, a concept of ownership that is antithetical and unimaginable to many Native peoples who live with the land and other-than-humans[2] in kinship (Salmón, "Kincentric"). It is grounded in the forced removal and genocide of Native peoples and their food sources and the enslavement of African peoples, the introduction of cheap and unhealthy commodity foods, and the unaffordability and inaccessibility of healthy foods for certain populations and neighborhoods. And as I will discuss, industrial agriculture and industrial food are grounded in multinational food and seed companies' battles over the language and definitions used to describe food (i.e., "cage free"; "natural"; "sustainable"; etc.).

Ecological community writing utilizes *distributed definition building*, through which individuals develop food literacy about what practices, labels, and definitions mean; about historical, social, and cultural issues connected to food; and about the systems they support when they choose certain foods over others. They generate their own definitions about these complex issues in ways that resonate most with them. Rather than try to control a term's definition and define it for an entire community ourselves or accept top-down definitions offered by those in positions of power, through *distributed definition building* community writing practitioners can redistribute power to community members themselves. I see my role, as a scholar-teacher-organizer interested in Writing and Rhetoric, as providing spaces for alternate definitions rooted in community desires and local knowledges.

Ecological community writing and *distributed definition building* move beyond a traditional community engagement partnership method in which an instructor develops a relationship with a community partner, determines ways to engage a class, and develops assignments that build a project or a

---

2   There are many terms used to explain what Indigenous cultures understand as the personhood of all beings, from trees to birds to stones. I have seen several terms from post-human, object-oriented ontology, anthropological, and new-materialist theories such as nonhuman (e.g., Gries, *Still Life*; Bennett; Sackey), more-than-human (e.g., Abram), and other-than-human (e.g., Blanco-Wells; Lien and Pálsson). In Indigenous scholarship I've seen more-than-human and other-than-human (e.g., Kimmerer, *Braiding* and "Serviceberry").

product for that partner in the span of a semester. They also expand on important established community writing theories of valuing knowledges of a community partner and the people they support (Green; Rousculp; Shah). *Ecological community writing* invites Rhetoric and Writing practitioners to intervene in community issues by building interdisciplinary, intercommunity, multisemester or multiyear coalitions to help communities write their own definitions of complex and contested social issue terms. It is a methodology based on an ecological, saturation approach that extends beyond a single moment, a single product, a single instructor, or a single discipline.

Local Organic moves between conversations in many disciplines including community-based writing and community literacies, food studies, Indigenous studies, ecological and new materialist rhetorics, ecopsychology, community engagement/service-learning, and systems thinking, offering an investigation into writing's function as an ecology-building tool within a particular place. In this sense, as many Rhetoric and Writing Studies scholars argue, writing is a part of an ecology. But it is more than that. Writing is also ecological. It is the distributed, networked, circulated, and embodied connection and co-creation between things in that ecology (Dobrin, *Ecology*; Edbauer; Ridolfo and DeVoss; Seas). And, I would add, *our work is ecological*, and viewing it as such expands community writing's potential. Based on years of teaching and project-building with students, farmers, government officials, activists, and other community members, *Local Organic* theorizes how our work to encourage distributed writing becomes an ecological thread connecting people, things, and ideas. Our intervention can be to help build the ecology itself through writing projects and events. That is *ecological community writing*.

When I use "we" and "our," I specifically write for Rhetoric and Writing Studies teachers, scholars, and graduate students who teach or research in communities, though I hope that if you are in another discipline, are staff working with community engagement, or are community organizers and activists, you will be able to adapt the ideas I offer in ways that best work for your areas of expertise and specific locales. I hope too that if you are a colleague in Rhetoric and Writing Studies who has never worked with community project partners in courses or scholarship, you will find inspiration in these pages, whether to try out a single community-engaged assignment or unit or to imagine a full course, course sequence, or research project. Parts of the methodology I offer can be used in theme-based courses and undergraduate or graduate research courses without a community-based component or in community organizing that is not associated with a university course.

A Critically Framed, Abundant Approach to *Distributed Definition Building*

Boulder County's stories of resilience and struggle are in many ways representative of larger national conversations about consumer confusion over food labeling and production practices, and the struggles of farmers to stay in business, as well as about food access, justice, and sovereignty initiatives that ask community members to acknowledge the violent history that led to current agricultural practices and to move into action accordingly; that acknowledge the connections between climate and environmental justice and food security; that develop and are led by members of communities affected by racism, ableism, and anti-Indigenous/Native practices within the industrial food system. Because community food security and resilience are some of the most important issues of our time—issues that tie to topics such as climate change, racism, poverty, disease, colonization, cultural sovereignty, gentrification, hunger, and language—community writing practitioners are well positioned to help launch initiatives in partnership with local communities and with faculty, staff, and graduate students across disciplines to address community-centered food literacies.

These topics can be overwhelming and depressing, causing people to turn away from engagement or to be consumed with anxiety. While we must acknowledge the issues named so far and the systemic nature of food-related problems, focusing attention on an abundant, relational, regenerative local food system can deepen a person's sense of agency necessary for these difficult conversations. *Distributed definition building* allows people from different ages, backgrounds, and positionalities to collectively "write" their community's food identity through activities and events at schools, parks, libraries, museums, and other public spaces.

Distributed definition building should be framed critically, aligned with an antiracist and decolonial approach to food justice. Through a critical, justice-oriented lens, *distributed definition building* urges an alternative to industrial food rhetorics that are premised on efficiency, exploitation, uniformity, and extraction. *Distributed definition building* embraces cultural sovereignty and foodways that have been condemned or marginalized, as well as Indigenous knowledges based in relations and kinship. It can tap into the deep yearning that so many humans feel for connection to the earth and other beings; for community; and for transparent, ethical food choices that keep us, our families, our cultures, and our communities healthy.

In projects with partners and students, I encourage an ethic in *ecological community writing* that is collaborative, imaginative, and celebratory in practice and action. With antiracist and decolonial framing, the abundance that *distributed definition building* offers does not mean promoting a culture of accumulation and hoarding of wealth and resources. It does not mean accepting an uncritical culture of toxic positivity or rejection of grief and anger about serious issues. Rather, as I will develop in later chapters, a critical approach to abundance offers an embrace of the complexity of emotions, a wholeness, a capaciousness, that allows for fear and anger about our food system to be acknowledged alongside joy, celebration, cultural reclamation, and gratitude as we move in action.[3] *Distributed definition building* also makes use of the abundance of skills and knowledges inside and outside of academia while moving us toward justice, accountability, and transformation. When *distributed definition building* is embraced through a critical lens, it urges a movement away from systems of injustice bound to industrial food production and toward reconnection with land, with animals, and with community. I have found that when my partners, students, and I encourage a critical and joyful relational approach to *distributed definition building* in community work around local food, we have been more successful in helping people to at once become aware of and articulate their desire for connection and relationship and to feel agency to lean into that desire.

Local Organic as an Ecological Text

Distributed definition building in the work of *ecological community writing* makes connections to and between the vast knowledges, expertise, and skills available to us. *Local Organic* responds to a vital need in higher education writ large to break down disciplinary silos and hierarchical structures based on rank and competition, to embrace and share knowledges

---

3 For a beautiful description of the embrace of wholeness toward collective healing and liberation, see adrienne maree brown's *Pleasure Activism*. brown argues that the erotic, the pleasurable, the political, and collective movement building are intertwined.

inside and outside of academia, and to encourage community-based learning and action. As a demonstration of abundant community-based knowledges, throughout the book you will find the insights of several farmers and local food activists and advocates in Boulder County from interviews I conducted with them between May 2020 and October 2021. These semistructured interviews, which lasted 60–120 minutes each and were conducted in person or over Zoom, help to map out an ecology of work happening locally to enact resilience and community food security. I had some degree of relationship with all of these people, most of whom I worked with for years, and who had been paid guest discussion leaders with my classes and partners on a variety of projects. I came to each interview with prepared topics I hoped to cover, but the conversations and questions evolved organically based on the direction the interviewee took in their responses. I recorded and transcribed each interview and shared it with the interviewee to edit or expand for use in this book. I cite them as local experts accordingly.

The incorporation of their perspectives embodies my understanding that I am only one small part of an expansive and dynamic ecology of people doing extraordinary work. It is our collaboration, relationships, and partnerships, more than any singular work, that matter. The incorporation of local experts' perspectives embodies the book's premise of knowledge cocreation and sharing, meaning making, and food literacy: that abundant knowledges, skills, and expertise are housed throughout our communities, not only in books, peer-reviewed articles, and universities. Some of my partners did not know one another or of each other's work until I began to see my role as a connector, *my work as ecological*, and that when we could collaboratively construct and share work and ideas across the local food ecology, our work would be more impactful. *Distributed definition building* allows for knowledge co-construction, connection, and celebration. *Ecological community writing* is justice based and active. It includes threads that lead back historically; it can cut certain threads of power and harm and generate new ones of accountability and repair; it can illuminate and account for gaps in who has and has not been included in the ecology. I imagine it as an always evolving, changing, dynamic system as we make connections between language, affect, things, and people, many of whom have been excluded from limited scholarly conceptions of knowledge sources.

In addition to the peer-reviewed scholarship, government and nonprofit documents, journalistic sources, and community expert interviews I draw from throughout the book, I also share information and analyses that I developed in situ to inform my engagement work. This set of data does not come from a single, cohesive study but rather grew in different ways at different times and for different purposes and audiences. For example, I coded by thematic area and analyze hundreds of interviews, community discourse analyses, and canvassing responses that my students gathered over several semesters, deepening my understanding of discourses about local food and developing methods for intervention as I went. This book incorporates this research and is based on years of experiential meaning-making with students and community members through undergraduate writing research, action-based research, and public projects. In the chapters that follow, I share artifacts, writing projects, and stories from my partnership work to illustrate *ecological community writing* and *distributed definition building* in action.

The *ecological community writing* methodology involves figuring out current conditions and who/what are involved. Readers wanting to use this methodology can create scaffolded course assignments (that may span semesters or years), mentor students as they conduct public research, and facilitate connections across an ecology of actors who may or may not yet know one another. In my own case of food studies, this occurred through a series of projects that are all parts of the *ecological community writing* methodology: research about the issue at a national level, followed by my students' interviews with restaurateurs, discourse analyses of local media, canvassing at the farmers market, connections to faculty and staff across campus to create coalitions, interviews with farmers and stakeholders, mapping of organizations and people in the county, and creation of events as spaces for *distributed definition building*.

Through the lens of my work over more than a decade with farmers, activists, and other local food leaders in Boulder County to understand what "local food" means and how to activate community engagement around that question, I look into types of ethical practice that depend upon reciprocal collaboration. Drawing on foundational theorists in community engagement, Rhetoric and Writing Studies, and food studies, as well as on the insights of community members, writing students, government organizations, and activists, my relational methodology centers community voices in a process of distributed writing, affirming that knowledge construction can be experiential, participatory, and asset based. This concept of distribution shapes my work developing writing opportunities and events across communities to incorporate knowledges from inside and outside of academia and across fields of study.

This ecological approach that incorporates many different genres of sources also expands narrow academic conceptions of what counts as expertise. It moves beyond the siloed, hierarchical, and exclusionary traditions of knowledge production and valuation so prevalent in colleges and universities. The book's structure is aligned with this philosophy of breaking down barriers through *distributed definition building* as I weave multiple disciplinary conversations, student and classroom work, community work, and my own reflections on the work into every chapter. Through an ecology of voices and ideas that I am helping to connect, *Local Organic* embraces the wisdom and intelligences that surround us. *Ecological community writing* offers Rhetoric and Writing practitioners working not only in food literacies but with a wide variety of emotionally and rhetorically complex social and environmental issues a methodology to guide engagement and literate action in our classrooms and communities by offering ways to publicly investigate and generate ideas and then help them to circulate.

How I Came to This Work

For thirteen years, I served first as associate faculty director for service-learning and outreach and then as faculty coordinator of community-engaged writing in the writing program at the University of Colorado Boulder until I left the university in 2021 for a position at University of Denver. As founder of CU's Writing Initiative for Service and Engagement (WISE) in 2008, I developed the first community-engaged writing courses for first-year students at the university and helped coordinate the writing program's transformation into one of the first writing programs in the United States to integrate community-engaged pedagogies throughout its lower and upper-division curriculum.

During that time, I offered professional development and pedagogical consultation to faculty and graduate students teaching first-year composition, topics in writing, science writing, business writing, and technical communication. Participation was optional, as not all courses are conducive to community-engaged work, and not all faculty or graduate students want to teach community-engaged courses. Facilitating faculty retreats, workshops, and individual conversations for those who were interested, every year I worked with about thirty writing faculty and graduate students to design WISE courses that enrolled over 1,200 students who in turn contributed over 15,000 hours of community-based writing work.

Soon my colleagues also led workshops on community-engaged writing, presented that work at conferences, and developed exciting community-engaged courses that influenced my own. We relied on, shared with, and celebrated one another. Liaising with several dozen community partners, I facilitated community-based writing workshops and collaborative discussions with them and our writing faculty over lunches and coffees as we determined how to engage in community projects and conversations through curricular intervention (House, "Community"). My colleagues and I wanted to help our students understand how people work on and discuss public issues within our communities and how Rhetoric and Writing impact and intersect with that work.

In my own writing courses, students balance theory and research with analysis of public discourses about food, analyze correlations between rhetorical concepts and community experiences, and learn to understand the complexity of arguments and the affordances of different genres for a range of academic and public audiences. My students, project partners, and I learn together, delving into complex, rhetorically informed conversations through inquiry and analysis and developing creative ways for people to become involved through writing. Community project partners make suggestions for readings and assignments to include in my syllabus and visit my classes as paid guest discussion leaders, students visit their farms and organizations, and my students and I have presented our work to government agencies, nonprofit boards, and other food-focused classes across campus.

Several years ago, inspired by ecological writing studies and systems thinking around "wicked" problems, I moved beyond the design of my own courses and my work with writing faculty on their individual courses, to help develop local food studies across the writing curriculum and across the university. Through a generous $24,000 grant from CU's Office of Outreach and Engagement and collaboration with faculty and graduate students in Communication and Environmental Studies, I launched the Dig In! project to facilitate interdisciplinary and community-based work. In collaboration with our community partners, we created undergraduate courses and graduate research projects around local food literacies, food systems, and food justice. I hope that this programmatic and interdisciplinary work that I discuss in detail in this book will inspire others interested in building faculty/staff/student cohorts or interdisciplinary and multicommunity coalitions through *ecological community writing*.

Writing programs, faculty, students, and staff, of course, function not only within institutions of higher education but within the town, county, state, and region they inhabit. Distinctions between town/gown and community/university are often arbitrary and can reify false hierarchies of knowledge production. Recognizing, respecting, documenting, citing, and connecting nonacademic knowledges are critical to ethical community-engaged writing work and are central to *ecological community writing*. Writing curricula and community writing projects will better serve students and communities if they organically evolve with that local, abundant ecology in mind.

Community Writing as Evolving, Relational Work

While the term "community writing" was sometimes used in scholarship and practice before I and my colleagues Alexander Fobes, Gary Hink, Catherine Kunce, Christine Macdonald, and Seth Myers launched the biennial Conference on Community Writing[4] in Boulder, in 2015, the term is now more commonly used in scholarship, course titles, and job ads. It is not always clear, however, what "community writing" means (House et al.). Variations on and elements of the concept gained traction in the mid-1990s as "composition's extracurriculum" (Gere), "community literacy" (Peck et al.), "literacy across communities" (Moss, *Literacy across Communities*), and later in that decade as service-learning, community service writing, and community engagement (e.g., Adler-Kassner; Bacon; Cushman, "Rhetorician"; Deans, "English Studies" and *Writing Partnerships*). Since the late 1990s and early 2000s, the scholarship and practice have developed significantly with key concepts such as "writing beyond the curriculum" (Parks and Goldblatt); writing about, with, and for communities (Deans, "English Studies" and *Writing Partnerships*); literacy, archival research, and historical work (e.g., Epps-Robertson; Royster, *Traces of a Stream*; Carter; Pauszek);

---

[4] I want to thank Seth Myers, Alexander Fobes, Catherine Kunce, Christine Macdonald, and Gary Hink for their tremendous work in helping me to imagine and host the Conference on Community Writing in 2015 and 2017. I also want to clarify that while not in wide circulation, the term "community writing" had been used by scholars before we launched our conference. The difference as I understand it is that the conference's term "community writing" serves as a broader and continuously evolving term that pulls together various strands of the field (see House et al.).

intercultural inquiry (Flower); community publishing and writing by and as the community (e.g., Kinloch; Mathieu; Mathieu, Parks, and Rousculp; Monberg); ethnographic research (e.g., Cushman, *Cherokee*; R. Jackson and DeLaune; Moss, *Community Text*; Roossien and Riley Mukavetz); community engagement in writing program administration (e.g., House, "Community" and "Keep Writing"; Rose and Weiser); community literacies (e.g., Baker-Bell; Feigenbaum, *Collaborative*; Flower; Goldblatt; Grabill, *Writing*; Long; Parks; Pritchard; E. Richardson, "She Ugly"); public rhetorics (e.g., Flower; Gries, *Still Life*; Long; Ryder; Rice); the public turn (Farmer; Mathieu); the political turn and Writing Democracy (Carter and Mutnick); decolonial and antiracist pedagogy and research involving communities inside and outside of academic spaces (e.g., Alvarez, *Brokering* and "Taco Literacies"; Baker-Bell; Cushman, *Cherokee*; R. Jackson and DeLaune; L. King et al.; Kynard, "All I Need," "Oh No," "Teaching," and *Vernacular*; Maraj; A. Martinez; Ore, *Lynching*; E. Richardson, "She Ugly" and "Coming From the Heart"; Ruiz); and the development of related journals such as *Reflections* (2000), *Community Literacy Journal* (2006), *Literacy in Composition Studies* (2013), *Spark* (2017), *The Journal of Multimodal Rhetorics* (2017), *constellations* (2018), *Latinx Writing and Rhetoric Studies* (2020), and *Rhetoric, Politics, and Culture* (2021). I understand "community writing" to embrace and continually evolve with these diverse areas (House et al.).

It is not that "community writing" is so broad a concept that it means everything and nothing all at once. Rather, I believe that sometimes the naming of subfields and fields of study can be designed either to reenforce an us/them binary that is exclusionary or to claim intellectual territory and "own it." Academia, Rhetoric and Writing Studies included, is premised on the claiming of ideas, on branding that idea in order to get ahead professionally. The academic drive to claim knowledge and to invent "new" ideas can sometimes make for a toxic culture of competition and scarcity. Historically and continually, it also has ignored ideas and scholarship from BIPOC scholars. I'd rather be part of a collaborative group of people who work to build one another up rather than tear one another down; to share, enhance, and challenge one another's ideas with generosity; and to acknowledge, celebrate, and be accountable to the intellectual and cultural lineages that are meaningful to us.

This of course doesn't mean that scholars should stop writing about theoretical ideas, and it doesn't mean that everyone should teach community writing courses or engage in public scholarship. But it does mean that we value and prioritize community-based scholarship and writing as an essential component of disciplinary work. Specifically, I'm referring to scholarship about community-engaged work and our own public writing products that are frequently misunderstood as less intellectual and rigorous than pure theory or as service rather than scholarship (House, "Re-Framing"); I am also referring to collaboratively written work (often with students or community project partners) that is sometimes counted in promotion and tenure cases as less significant than single-authored work—reinforcing ideologies of individualism and isolationism—when community-engaged work's cornerstone is reciprocity and collaboration. As in agricultural polycultures, where different plants and animals have different but essential roles and work together for the health of the whole, I strive to participate in a knowledge-sharing and knowledge-generating polyculture (S. Wade) based on antiracism, access, humility, mutual mentorship, and generosity to build success for the whole ecology.[5]

Many of the foundational texts in community writing foreground ethical considerations of the university's or college's engagement with the broader community and the need for reciprocal partnerships in which faculty, community members, and students each gain from and give to the relationship. Paula Mathieu was one of the earlier engagement scholars in Rhetoric and Writing Studies to express important concerns about the ethics of service-learning, which, she noted, seemed to be trending toward "long-term, top-down, institutionalized service-learning programs" that can "burn" community partners, who "don't live or think in terms of semesters or quarters or final exams or spring break" (96, 99). adrienne maree brown reminds organizers, in a play on Stephen Covey's concept, to "move at the speed of trust" (*Emergent* 42). In this vein, Eli Goldblatt's view of writing program administrative work is premised on relationships coming first: "[W]hat if we start from the activist's ground in this instance, learning before we act, developing relationships and commitments before we organize classes and set up research projects?" (130). Respect and relationships are a foundation of community-engaged work (e.g., Arellano et al.; Blackburn and Cushman; Rousculp; Goldblatt; Hsu; M. Powell; Shah). Relationship-focused projects require faculty and students to decenter

---

5 The Coalition for Community Writing offers a "Promotion and Tenure Resources."

themselves and to understand that the reciprocity, rather than any grade or scholarly outcome, is primary.

Carefully built and maintained relationships require significant time beyond the time spent on classroom-based course planning and, as Mathieu says, cannot necessarily be bound by semesters and academic calendars. Jo Hsu has encouraged critical questions about this issue:

> Academic Time™ refuses to account for human bodyminds and community relations—and . . . discourses of scarcity and precarity are used to enforce exclusionary timelines. One of the major challenges of working in cultural rhetorics is trying to hold our relationships in balance: How do I prioritize the relationships and the actions I find meaningful while also keeping my job (which, practically, enables me to make those relationships and take those actions)? (in Arellano et al.)

Pressures on faculty and graduate students to publish quickly and often are at odds with relationship-focused timelines. All too often institutional mission statements that affirm commitment to community do not align with realities of support or with institutional policies around review, tenure, promotion, and hiring priorities for community-engaged graduate students and faculty. While individual faculty and graduate students may be committed to the intensive work and time involved in community-engaged teaching, public writing, and scholarship, people in the position to help make change and offer support can work toward structural programmatic and institutional change. The answer should not be to warn graduate students and junior faculty to wait to do community-engaged work until postgraduate school or post-tenure, as is too often the case. Rather, it is essential to change the systems and mindsets that make such meaningful and important work so difficult (Taczak, House, and Carter-Tod).

Practitioners have responded to important logistical and professional concerns in the form of methods and practices to work toward reciprocity and collaboration so that the community project partners are not unnecessarily burdened and instead gain benefit, while well-designed community-engaged projects can enhance our students' learning and enrich our own professional lives.[6] Scholarship from the last decade takes up the challenges

---

6 The Coalition for Community Writing, which I founded with a national cohort of scholars in 2018 and for which I serve as executive director, is an organization committed to systemic changes in institutions, programs, and policies, and offers conferences, workshops,

set forth by community writing's earlier scholarship to articulate the ethical and logistical as well as antiracist, anti-ableist, anticolonial necessities for reciprocal community-based work and writing *as* community (Baker-Bell; Hubrig; Cedillo; Hsu; Monberg) that is asset-based rather than deficit-based (e.g., Anderson; Bernardo and Monberg; Del Hierro et al.; Opel and Sackey; Parks; Roossien and Riley Mukavetz; Shah). As Keisha Green explains that as an antiracist approach, "asset-based theory means focusing on what already exists, what is already happening in the context, builds on what already exists, acknowledges what is present" (155). Elaine Richardson's work with Black girls' and Black women's literacies (e.g., "She Ugly"), Beverly Moss's ethnographic research on churches and church sermons (*Community Text*), and Ada Hubrig and Christina Cedillo's work on disability justice (e.g., "Access as Community") offer examples of scholars embedded in their communities doing research with and as members of the community themselves.

It can pose an ethical problem when a practitioner who is not yet known or trusted by a community tries to enter with a project in mind. Cana Uluak Itchuaqiyaq emphasizes that an essential role of community-based scholars who work as members of their home communities is to restrict access to that community from scholars who may exploit it for personal gain. These "boundary spanners, . . . individuals who occupy both academic spaces and marginalized community spaces and who are called on to act as mediators between the two," Itchuaqiyaq explains, can help ensure that research and project ideas "related to Indigenous land and peoples must come from these communities themselves (97, 98). As scholars, we need to respect community sovereignty and be humble enough to take the time to build local relationships and listen to local needs and wants and pivot our existing research and restructure new research questions to help fulfill those needs" (97, 98). Respect and humility are essential in forming relationships. A mantra from disability justice that has been adopted by many marginalized groups is "Nothing About Us Without Us" (Hubrig). Careful work and humility require our understanding of when to decenter ourselves, when engagement is not wanted, or when to commit to the public writing that is needed even if it will have little to no professional "value" in the academy (L. Smith).

---

resources on its website, and mentorship for making individual and infrastructural arguments for community-engaged teaching and scholarship.

Relational work requires deep, contemplative personal work as we interrogate our own and our ancestors' relationships to systems of oppression and legacies of violence (House and Briggs). This personal work and accountability become part of the ways in which we relate to and define our ecology, whether that be a family, neighborhood, region, department, or discipline. It can help us understand how we can (or should not) interact with various parts of our communities or, especially, with communities of which we are not members. For this reason, I share a relevant part of my ancestral connections and ideological legacies in "On Lineage and Purpose" and other places throughout the book. When working toward a just local food system and a just, locally focused curriculum, relationships—past, present, and future—and the lineages (be they familial, cultural, linguistic, disciplinary, or connected to foodways) they foreground or suppress are essential to consider in our course, research, and partnership design.

When community-engaged Rhetoric and Writing Studies practitioners discuss relationships and reciprocity, the literature has typically referred to a human triad of community partners, faculty, and students. Many Indigenous activists and scholars across disciplines remind us, however, that knowledge emerges from the land on which it is created and from the plants and animals who share that land with us (e.g., Kimmerer, *Braiding* and "Serviceberry"; LaDuke, *All Our Relations*; Roossien and Riley Mukavetz; Ríos; Simpson). The local food work happening in Boulder County is deeply interwoven with the history of the land and conflicting ideas of ownership and kinship. Andrea Riley Mukavetz writes about the primacy and complexity of relationships, including with land, in her ethnographic coauthored book: "the kind of relationships one has with family, with community, across generations with trauma and recovery, with ancestral land, with writing and research, and so on" (Roossien and Riley Mukavetz 11). In a time where food insecurity, dispossession, culture loss, and separation from our other-than-human collaborators appear to be perennial and life threatening, ecologies of local food, and definition building about local food in those ecologies, must include history, ancestors, land, plants, animals, water, and others, opening ourselves to their rhetorical agency and knowledge (House and Zepelin). These Indigenous concepts have profoundly guided my work to reconceive the traditional community writing partnership model into an *ecological community writing* methodology that includes other-than-human partners, as I discuss in the final chapter of this book.

Local Organic offers a methodology that decenters students, faculty, and universities and colleges themselves, to function as parts of a vast ecology of relationships. The concept of a university or college as part of a place or network or web is not novel. When Goldblatt asks, "Is there something about the demography, geography, or social psychology of a region that should affect the instruction and investigation pursued inside a given campus?" (11), the answer of any community-engaged scholar is, surely, "yes." Like Goldblatt, Stephen Parks gets at the concept of a writing program as part of a dynamic web: "[T]he possibility had to be created for participants to come to see the ways in which their personal viewpoints were simultaneously part of a web of existing or emerging progressive community-based institutions" (33). My approach grows from this previous scholarship to propose an expansive way to conceive of our place and work in an ecology.

Why an Ecological Approach to Community Writing Is Useful

If building reciprocity and care in relationships is key, then listening to communities across different perspectives, needs, and cultural paradigms proves critical as well, in terms of developing those relationships and ensuring broad consideration for who is at the table in community discussions about definition building. Jeff Grabill encourages a "listening stance" (*Writing* 124), which provides people from diverse backgrounds and perspectives "the rhetorical ability to participate effectively and the structured requirement to listen to what others say" (124). Romeo García reminds us that "community listening encourages listening for humanity in stories and memories in between cultures, times, and spaces" (7). Like many scholars have argued, Paul Feigenbaum (*Collaborative Imagination*) reminds, "[i]t is especially important that people silenced by traditional institutional structures participate robustly in the inventive processes" (66). Rachael W. Shah's *Rewriting Partnerships* argues for the inclusion of perspectives on the benefits and outcomes of community-engaged work, not only of nonprofit staff but of community residents themselves, in our scholarship. Shah maintains that inclusion of these "scholars off the printed page" (13) is essential in any ethical assessment of the efficacy of partnerships and products of those partnerships. We have to make sure that their inclusion is what they want, that it serves them well, and that it is not exploitative. Itchuaqiyaq writes, "[Accountability] means caring more about the needs and safety of my community more than my own professional needs" (99). It's not just a matter of getting people to the table—it is the collaborative invention once they are there that leads to action. Reciprocity means that we are accountable to one another.

Distributed definition building is a method for how to enact these theories in on-the-ground community writing that encourages communities to become involved in definition building around critical terms, definitions that can impact the writers' lives and the lives of others in their community based on who has the power to control those terms. Beyond those in positions of organizational power, this method also challenges us to consider who in the community is showing up and asking to be heard. And then it is also a matter of how we listen. In their theorization of community listening, Jenn Fishman and Lauren Rosenberg explain, "[C]ommunity listening is about being immersed in the experience of understanding and non-understanding, trying and trying again with empathy. The listener is in a position of generous openness. From this stance, it becomes possible to pay ongoing, unflinching attention where it is needed most" (3). In that spirit of unflinching attention, we can also listen for the gaps. Given that there are inevitably groups missing from community conversations, how might rhetoricians or writing faculty and students develop possibilities "to hold all accountable and to carry on in the name of justice" (García 12)?

The community that I am both studying and a part of—the people involved in building the local food movement in Boulder County—is one I have helped to organize around ideas, and of course so many in this community have helped to organize my work around their ideas as well. As adrienne maree brown writes, "[s]ome of us are trying to imagine where we're going as we fly. That is radical imagination" (*Emergent* 21). Because the community actors and conversations are always in flux, and because my own understanding continuously grows, my work is always evolving, focused on developing food literacy through community-based courses and projects involving students, faculty, community leaders, and other members of the public in order to inquire into the definitional nature of "local food," to collaboratively problem solve, and to build together something meaningful and more focused toward access and justice.

While I offer a usable method and examples in this book, part of the relational methodology of *ecological community writing* is in recognizing its iterative and locally specific nature. For this reason, I've deliberately written this book to blur lines between theory and practice. When I offer my work, I discuss how the ideas evolve continually, how students' projects help shape the ideas, how shifting partners shape the ideas, and how place itself shapes the ideas. I've woven theory, pedagogy, narrative, and community-based work through every chapter with a belief that they are so intertwined that I cannot separate them. I hope that you will be able to adapt and continue to develop the methodology and method based on your own community's conversations and priorities in ways I'd never have thought of. We're imagining as we fly.

In a Zoom interview for this book that I had with a former writing student, Madeline Nall, who went on to work as communications manager for five years at a farm they first visited during our class trip many years ago, they explain the importance of *ecological community writing* in our class in contrast to what can often happen when students do not engage in community conversations: "I think universities often feel alienated from the communities in which they exist because there are so many transplants, like myself, who don't have existing ties and spend most of their college careers trying to integrate with the college culture rather than invest in the community." They continue, "[O]ften when these students graduate, they take their degrees and move elsewhere, after having likely participated in the gentrification of the city that made that degree possible." When colleges and universities "incorporate and center community voices and connections, like we did in your class," Madeline says, "then students would feel more invested in supporting existing community work or contributing new perspectives to further that work."

I ask Madeline what kind of a shift this ecological approach to engagement may cause in students' feelings of agency and connection or in their research trajectory. They do not hesitate in their answer: "I think this would help drive a paradigm shift in terms of where and how systematic change is best leveraged, further empowering students to be change-makers, as they can learn from leaders directly and witness the power of strong community ties." Madeline signals to the shift that they felt when our class moved from reading and researching about food issues to visiting farms and dairies in Boulder, and they get at the importance of *ecological community writing* that is central to this book. Madeline explains, "There are so many cool movements happening, and it's also exposing this great divide with people like academics who know the code and know the language and have studied these things and know theory yet are so removed from the communities at stake. Community leaders hold just as much value and knowledge as experts in academia, and universities should prioritize building those partnerships." Indeed, community leaders are essential in the distributed definitional writing of what "local food" can mean, how people understand the systems they support through their purchases and practices, and what is possible to produce in Boulder County.

While students create individual and group writing projects in my classes, I emphasize the collaborative, ongoing, and circulatory nature of writing; its embeddedness in community; and the ecology of people, organizations, and things that continuously shape our writing and write with us. As Madeline describes the importance of this distributed writing method for undergraduate students, I would suggest that it is equally important for faculty and graduate instructors. Our community-engaged curriculum may indeed change our community, and our specific communities and places themselves can also change our curriculum, as Boulder County has changed mine. The places where we live are both a physical place and a rhetorical idea, as "local food" is both physical thing and rhetorical idea. As I will describe in detail throughout this book, this understanding has changed my teaching and scholarship as I adapt accordingly.

In contrast to the problematic power of certain institutions and companies to define what "local food" is or should be in Boulder County and elsewhere, the projects I help to coordinate with students and partners offer distributed power to communities to define the term for themselves based on their values. I do not anticipate that a single definition for local food will (or should) emerge but rather that the writing projects and events will spur inquiry, food literacies, and emotional connection in individuals, who may then use their new knowledge in their eating decisions, and that these activities may, in the long run, prove to impact others' actions and even local policy. The purpose of my community work is not to reach consensus about what local food's definition should be or to universalize desire for it. It is to put lots of ideas on the table, so to speak, in the spirit of inquiry, and to listen for what emerges. But I don't believe that our work should stop with the facilitation of inquiry. "In short," as Linda Flower notes, "the meaning of a local public lies in what it *does*" (42, emphasis original). When my partners, students, and I promote critical food literacies, I consider what actions these literacies enable and encourage people to do and how our work might lead people toward visions of collective power and justice.

Critical Food Literacies and *Local Organic*

As facets of the interdisciplinary field of food studies, critical food literacies and food rhetorics are increasingly visible educative endeavors among Rhetoric, Writing, and Communication scholars. This is evidenced in panels, workshops, and institutes at major conferences such as Conference on College Composition and Communication, Rhetoric Society of America, and Conference on Community Writing; academic books and journal articles too numerous to be inclusive here (e.g., Alvarez, "Taco Literacies"; Boerboom; Brewster; Broad; Conley and Eckstein; Donehower et al.; Dubisar; Frye and Bruner; Goldthwaite; House, "Keep Writing" and "Re-Framing"; Martinez; Schell et al.; S. Wade); journal special issues dedicated to food studies theory and literacy (*College English*, vol. 70, no. 4 [2008]; *Pre/Text*, vol. 21, no. 1–4 [2013]; and *Community Literacy Journal*, vol. 10, no. 1 [2015] and vol. 14, no. 1 [2019]); and undergraduate readers from Bedford, *Food Matters* (Bauer), and Fountainhead, *Food* (Rollins and Bauknight). Critical food literacies incorporate issues of racial justice, anti-ableism, and cultural sovereignty as well as reclamation of culturally specific practices and a decolonization of our diets (e.g., Alkon and Agyeman; Alvarez, "Taco Literacies"; Garth and Reese; Simpson; Kimmerer, *Braiding*; Martinez; Mihesuah and Hoover; Penniman, *Farming*; Salmón, *Eating*; R. White). Critical food literacies teach people to interrogate the relationships between themselves, language, food, land, corporate capitalism, and white supremacist culture—living legacies of land theft and forced labor that built white cultural wealth and the current food system in the United States. *Ecological community writing* enacted through *distributed definition building* and the excerpts of interviews with community partners that I include in this book focus on these relationships, making connections between local discourses and the larger systems of which they are a part.

Ecological theories of writing necessarily shift what happens inside of a classroom and in our various communities. As I've developed the methods described in *Local Organic*, I've woven much contemporary Rhetoric and Writing scholarship into my praxis as a community writing scholar, teacher, and community member in order to develop my own version of an ecological methodology specifically for community writing. My hope is that *ecological community writing* will help you to catalyze writing for impact in your own communities while following the key tenets of reciprocity and asset-based, ethical engagement that are so critical to community-based practice and to Indigenous conceptions of ecological kinship (Shah; Cushman, *Cherokee*; Kimmerer, *Braiding*; Peña).

Local Organic demonstrates my ongoing learning with how community writing can build from a weaving together of theories, practices, and priorities—and how our own theories, pedagogies, and activism shift in response. Laurie Gries argues that rhetoric is not "still" but has life: "Once unleashed in whatever form it takes," she explains, "rhetoric transforms and transcends across genres, media, and forms as it circulates and intra-acts with other human and non-human entities" (*Still Life* 7). Rhetoric is more "like an event—a distributed, material process of becomings in which divergent consequences are actualized with time and space" (7). If we can enable distributed writing around the social, environmental, or cultural issues we're working with, then we're functioning not just at the level of individual writers anymore but also at the level of ecology. When we help to move private or marginalized voices into public dialogue, the words and ideas become part of systems of circulation that can sometimes have impact, whether on personal decision making, community decisions, or policy.

The theoretical strands named in the preceding text influence my study of "local food's" rhetorical life across genres and audiences in Boulder County. While I draw on national rhetorics of food literacy, I use Boulder County's food localization efforts as a localized example of engagement in which Rhetoric and Writing practitioners working with a wide variety of emotionally and rhetorically complex issues are well positioned to encourage literate action in classrooms and communities through a public process of investigation and action. This book offers a point of entry into current discussions in Rhetoric and Writing scholarship about critical food literacy and, more broadly, argues for the exigence for public writing students, scholars, and teachers to help build ecologically informed community writing coalitions. The distributed definitions of "local food" that I study in this book come from community writing at events I have helped facilitate, interviews with community leaders, community surveys, promotional materials, websites, policy documents, journalistic work, student writing assignments, peer-reviewed scholarship, and from the land itself.

How This Book Is Organized

Before moving into the chapters, I locate my own positionality in this project as a researcher and teacher working toward accountability and action in the section "On Lineage and Purpose: A Place to Begin."

In chapter 1, "Food Rhetorics and a Crisis of Definitions," I discuss a first stage of the *ecological community writing* methodology—research about the issue at a national (or global) level. In food studies, this means understanding the broad, systemic issues on a national scale and the connections between language and control of definitions, corporate versus community-based power, and some of the most serious issues associated with our industrial food system as they relate to language. I make the case that these issues are directly connected to the fight over definitions associated with industrial, organic, and local food production, distribution, and consumption, as well as food access, food justice, and food sovereignty. Control over definitions matters. Despite the high costs associated with cheap food, relatively few people living in the United States know about food production practices and their implications, the environmental and human health costs, or the contents of the food we eat. This illiteracy is a deliberate condition that multinational seed and food companies work tirelessly to ensure, often through confusing advertising that uses definitions for food choices that the companies have created in order to sell products rather than to inform consumers. Even fewer people understand how legacies of land theft, exploitation, and slavery are still present in our food system. As an alternative to models of behavior rooted in competition, obfuscation, and extraction, local food justice work—the work of access, equity, and asset-based models of community engagement—exposes and disrupts corporate and top-down definitions through local organizing.

Chapter 2, "What's 'Local' About Local Food? Rehinging Words to Meaning," moves from the national battles over definitions discussed in the previous chapter to the competing discourses about "local food" in Boulder County to model how definitional tensions around the term can lead to consumer confusion, misleading marketing, and, sometimes, stalled action. In this chapter, I explain how my long-term research project began as a community writing collaboration with one class that then organically spawned years of investigation and projects, ultimately leading me to develop the *ecological community writing* methodology and *distributed definition building* model that I offer in the book. Through rhetorical analysis of several hundred texts including interviews with Boulder County farmers and restaurateurs, stories in local media, and canvassed responses from farmers market patrons—all gathered by my undergraduate students through class projects—I reveal local contradictions and confusion over the assumed meaning of "local food" on the part of community members versus the realities of practice.

In the county there is active work across various sectors toward localization of the food system, but there are many different, conflicting, and sometimes deeply problematic definitions circulating for what "local food" means. I investigate the dissonance between the material and imagined realities behind Boulder County's local food rhetoric and delve into food and language as they intersect with consumer awareness and agency as well as farm viability. This study of the contested term "local food" highlights the need for *distributed definition building* that offers community-written, bottom-up definitions. The national and local research that I discuss in these first two chapters demonstrate the need for (and offer methods such as interviews, surveys, and discourse analysis that practitioners can use to toward) collaborative, interdisciplinary, intercommunity *ecological community writing*.

In my study of responses to local food movements and how activists and advocates can effectively communicate about the issues at stake, I have been influenced by several psychology studies that analyze the effects of different kinds of messaging about traumatic issues on behavior. In chapter 3, "Crisis and Abundance Rhetorics in Moving People Toward Collaborative Engagement," I focus on affect as an element of or an actor in an ecology. To do so, I study other elements necessary in *ecological community writing*—knowledge drawn from other disciplines—climate communication, ecopsychology, and trauma-informed research. Learning from these fields about the reasons humans reject certain messages even in the face of overwhelming factual evidence, I develop a lens through which to implement my method that may help shift people's responses. I explain the influence these findings have had in the development of *distributed definition building* as well as how they can aid in communicating about any number of issues to our students and to the public. In this chapter, I encourage practitioners to adopt the lens that I call *critical rhetorics of abundance*. In food literacy work, this may mean countering some of the common deficit and fear-based rhetorics around industrial food and food crises by creating positive messaging and embodied options that encourage involvement at the personal, community, and policy levels.

In the next two chapters, I delve into communication strategies, ecological writing theories, and systems theory to argue for how public rhetoricians can help communicate messages effectively as well as create opportunities for community members from different positionalities—ages, races, ethnicities, financial means, and backgrounds—to collectively "write" their own community's definition for "local food" that aligns with their values. Chapter 4, "Writing Abundant Ecologies: From Theory to Collaborative Local Research," offers an overview of ecological, systems, contagion, and circulation theories that have shaped my praxis as I consider further parts of an ecology—the human and other-than-human actors that contribute to the possibilities for local food's production, consumption, and appreciation in Boulder County, as well as to the rhetorical life of the idea of "local food." In this chapter, I offer strategies for visualizing the relevant work, people, organizations, and places in your own ecology and for building coalitions across campuses and communities.

Concrete examples of the *ecological community writing* methodology in use as well as how *distributed definition building* is generated in my courses, outreach to scholars and groups with similar interests, and community-building work are woven throughout chapter 5, "Ecological Community Writing in an Abundant Foodshed." In the definitional confusion around what "local food" means comes an opportunity for *ecological community writing* that offers ways for people to write about their relationships with local food and to experience its abundant nature. It is a way for community members to have a stake in shaping public discourse or policy discussions that often seem so out of reach for ordinary people. In this chapter, I weave theory with several examples of public writing events my students, project partners, and I have organized to demonstrate how Writing and Rhetoric scholars and teachers can apply ecological and systems thinking in their classes and community-based work to help identify and connect actors working toward a desired outcome.

These chapters offer specific examples of my food literacy work through the interdisciplinary and intercommunity "Dig In!" project I founded, which catalyzes public food research projects, writing courses, and public events at museums, libraries, parks, farms, and schools involving students, faculty cohorts, community leaders, and members of the public. These partners and participants collectively inquire into the complexities of the local food movement and collaboratively work toward a resilient local food system. Through all of this work, I theorize how writing is an ecological thread connecting people, things, and ideas.

In chapter 6, "Abundant Partnership: A Local, Organic Approach to Rhetoric and Writing Studies," I center Indigenous scholarship on reciprocity and animacy, as I consider how concepts of partnership, *ecological community writing*, and definition building deepen when we consider how other-than-humans partner with us in writing. I offer a theory of community partnership and suggest that an ecological perspective that includes other-than-humans as collaborators in definition building can shift our work.

In the following chapters, I demonstrate the chronological development of the two main concepts I offer in this book: the *distributed definition building* method in community projects and class assignments that led to the relational methodology of *ecological community writing*. I believe that transparency in this iterative development over many semesters and years is important, but any use of the examples of assignments and events that I offer does not need to be applied linearly. Especially in community-engaged work, relationships deepen or change, community conversations and new policies shift practices, students learn lessons that alter the trajectory of projects, nonprofit staff leave or join organizations, our awareness expands and transforms. That messiness is real and essential and can be generative. As I weave together the works, theories, and authors that have meant so much to my own development with direct dialogue, collaborative writing, development of events and projects with others in our community (e.g., in real time and asynchronously), and indirect study (e.g., through creation and analysis of different kinds of texts and publications), I theorize our understanding of ourselves as embedded in larger human and other-than-human systems as we also help to generate connections in those systems.

The projects I present are not only chronologic but sometimes happening simultaneously. The distributed and simultaneous nature of ecology-building work is hard to write about because it is *not* linear. I ask for some playfulness on your part in trying to visualize the work and theories I describe. There are multiple ways I've considered demonstrating a visualization, none of which is adequate. One way could be to imagine a constantly growing and connecting 3D model of an ecology with the contested term such as "local food" at the center, and emanating from that term are dozens of connections to language, histories, affect, humans, organizations, theories, projects, other-than-humans, and then dozens or hundreds of connections emanating from those terms to other things, and on and on. Each chapter in the book offers connections between "local food" and various parts of an always-in-flux ecology. Another idea could be to imagine each chapter as a transparent overlay that draws out different ecological parts, and whereby, as you overlay each sheet, the ecology becomes more connected and detailed. I don't know that there is a best way to imagine *ecological community writing*, so I encourage your creativity in visualizing.

I invite you to make connections with *ecological community writing* in your own teaching, research, and activism or organizing work. An anonymous reviewer, for example, suggested a connection that deserves further investigation between ecological community writing and undergraduate research in writing studies, particularly course-based undergraduate research experiences. Another colleague who read a draft of the book saw usefulness in the distributed definition-building method for her work with students to define "civic education." I welcome these and other connections to and extensions of the work I offer. There are myriad ways to use *ecological community writing*—in course development, faculty retreats, staff trainings, graduate courses, Writing Across the Curriculum workshops, and more. As the book's title, *Local Organic*, suggests, I hope that you will adapt and play with methods that can be used for *ecological community writing* in ways that develop organically through your experiences with them and that are most useful in your own locale. Some of the projects and research I describe can be done through writing courses without any funding. Other projects required internal grants from the writing program and the campus's engagement office but could be modified based on available resources. As I discuss how my methods and methodology developed, beginning with a brief introduction to myself and my positionality before the first chapter, I offer ways for how to listen to our community's unique conversations as we attune to community building and justice-oriented definitional work from the ground up.

On Lineage and Purpose

A PLACE TO BEGIN

When I accepted a faculty position at the University of Colorado Boulder, where I worked from 2006 to 2021, I thought I was moving to a place far removed from my Maryland and east coast roots. Taking the job in a place I didn't know was part of what accompanies the academic life for most recent PhDs who seek a faculty position. I was part of a workforce determined to take whatever job we are lucky enough to get, in whichever part of the country. Academics bring expertise from one place to the next, often with the assumption that what "works" in one part of their country or world will work in any other, whether that is an assignment, syllabus, or area of study. I was hired as a newly minted English PhD to teach writing courses about literature and mythology, which was my academic background.

But Boulder, *the place*, called forth a new focus. Over several years, increasingly, *it argued* for my focus to shift to the local food and farming work that I am now involved in through my teaching, scholarship, and community work, as I will explain in the pages of this book. Boulder urged, and I listened. Any number of issues could have resonated based on where I would move for a job, but food literacy around local food and agriculture, which are vital conversations in Boulder County, found me—food illiterate as I was when I first moved here. Through my participation in public events, conversations, and interviews with farmers, consumers, government officials, students, and activists, I continue to develop my local food literacy. How people understand, discuss, and define "local food" in Boulder County is at the heart of this book.

Would I be a food literacy scholar and teacher if I had moved to another place? Maybe not. I might have felt an urging from *that* place, guiding me toward different work because, I believe, the place where we live can call forth our work. As the title of this book suggests, our work can emerge and develop organically from our specific locale. This growing belief has led me to think about how place (and its material, historical, and embodied rhetorics) and university and college writing programs can connect. In graduate school and my early years as a faculty member, I found inspiration in scholars such as Beverly Moss, bell hooks, Tom Deans, Eli Goldblatt, Steve Parks, Terese Monberg, Paula Mathieu, Derek Owens, and others, in their arguments for writing programs that function within the vision for literacy in a particular place, for writing beyond the curriculum, for tactical interventions, and for writing about, with, for, and as local communities. Those early service-learning, community literacy, engaged pedagogy, and place-based theories resonated deeply, and I rooted into a theoretical space to call home.

* * *

As I packed up to move to Boulder, where I knew no one, I was surprised when my maternal grandmother, Sally Rathvon, told me the story of my great-great grandmother Marinda May Platt, who had been a homesteader in Boulder in the 1870s–1880s and had studied to become a schoolteacher in a preparatory course at the University of Colorado. Suddenly, there was a connection between me and an ancestral past I hadn't previously known, the university, and the land itself. This connection is fraught with a mix of simultaneous emotions. *A curiosity* for knowledge about my ancestor, about lineage, and about the strange set of convergences that led me to Boulder: Where was the homestead located? Was I teaching in buildings at the University of Colorado in which Marinda had taken classes? What was her life like here? What kind of a teacher did she become? *An ever-present grief and anger* as I search for family information and learn more and more about the family's presence in Colorado during the most brutal period for Native Americans in the state's history. What did Marinda as a young child witness and internalize from the adults in her life? What stories and ideologies did she learn as a girl and teenager that shaped her? Did she accept them, question them, fight against them? How did other members of the family perpetuate or resist the violences endemic in settler colonial culture? *A sense of commitment* to read and learn about current work by Native-led organizations and scholars. How might I support and learn in whatever capacities, if that is wanted of me? *A desire for accountability* and for understanding. What am I called here to do? adrienne maree brown asks herself a question that guides my own reflective work as I think about ancestral and cultural legacies in relation to disciplinary work and community-based work: "Where in my own life do I still persist in actions that presuppose my importance and supremacy, rather than accept my small role in our collective existence?" ("emergent").

I found the one-room schoolhouse that Marinda Platt and her brothers attended: the Valmont School. It is now protected as a historic site in east Boulder. I brought my elder daughter, three years old at the time, to search for it with me. When Marinda studied there in the 1880s, it was a different building, and it was rebuilt in 1911. From old photographs at the Carnegie Library for Local History, I see the building she attended and imagine her as a girl in her long, starched dress looking back at us across the tall grasses of the schoolyard. From preserved family photos and accounts, I know that her skin would be darkly tanned, matching her black hair that reached to just above her knees when down and her dark eyes, which her husband, Spencer, said he noticed first about her.

In 1859, the year gold was found west of Boulder, Marinda's maternal uncle Jerome Gould had gone to Colorado with a bull train, settling there for good in 1861. Her grandparents and several uncles soon followed, which means that they were in Boulder County at the time of the 1864 Sand Creek Massacre described in the City of Boulder Staff Land Acknowledgment with which I open this book. Marinda's parents joined the family in 1876 when Marinda was six years old, traveling from Iowa by covered wagon to live on the family homestead in what is now northeast Boulder (at the time, where they lived was called Valmont by settlers). Farming there, as she described it years later in a letter to one of her sons, was "the hardest kind of hard work."

Was I standing in precisely the spot where she once stood? My daughter and I leaned against the fence and locked gate that barred us from entering the school property. My girl said, "There are so many mothers. You're a mother, and grandma, and her mother, and her mother, and her mother." "Yes," I said, "So many women. And now you." I jiggled the chain on the fence, hoping it would open. I strained to see into the past.

There was something profound about showing my young daughter, born in Boulder, where her great-great-great grandmother had gone to school. Though I hadn't known of Marinda before moving to Boulder, as a girl I grew up on stories of the strong women on my mother's side. My great-grandmother had been a suffragette in Chicago; my grandmother was an ardent feminist, writer, and civil rights advocate. But as far as I know, legacies of settler colonialism were not questioned. For whatever reason, the family's legacy as farmer colonizers on unceded land seems to have been separate in my ancestors' minds from women's and Black Americans' fights for liberation. In an apparent disconnect between legacies of violence, my grandmother would tell me to be tough and resilient because, she'd say proudly, "You are of pioneer stock," a saying she had learned from her mother and passed on to my mother. I am proud of many things about my grandmother, but this is a narrative that I will not pass on to my daughters. The family legacy, the land, the place itself call me to accountability in ways I am only beginning to comprehend.

Robin Wall Kimmerer writes in *Braiding Sweetgrass*,

> America has been called the home of second chances. For the sake of the peoples and the land, the urgent work of the Second Man may be to set aside the ways of the colonist and become indigenous to place. But can Americans, as a nation of immigrants, learn to live here as if we were staying? With both feet on the shore? ... What happens when we become truly native to a place, when we finally make a home? Where are the stories that lead the way? If time does in fact eddy back on itself, maybe the journey of the First Man will provide footsteps to guide the journey of the Second. (207)

Kimmerer's questions for white settler colonists are critical. I read her as urging a shift of mind and of being that are, I believe, essential for the survival of the human species. We can consider these big questions on a global scale, a local scale, and an individual scale.

Scholars, students, and teachers can find new ways of being in the academy that are interwoven with place so deeply that the content of our scholarship and teaching are fundamentally altered by where we live. *The place calls forth our work*. In my teaching, writing, and advocacy with food studies, I can ask what the current stories being told here in Boulder County are about cultural heritage, farming, and the history of this land. I can analyze how language choices reinforce or "set aside the ways of the colonist." I can study the pervasive stories told in this area about food, consumption, and individual success, as well as help to imagine with my community different stories that might lead to a new way of being for me, for my daughters, and for others. I need to reflect deeply on ways in which the indoctrinated stories of white supremacy culture affect my understanding, the mistakes and missteps I will inevitably make, and the obligation to be accountable. These issues are at the heart of this book and in the connections that I draw between food, language and definitions, *ecological community writing*, and power.

* * *

Marinda's family bought her maternal uncle's "land claim," growing hay and raising livestock. Her uncles had claimed property across dozens of acres of surrounding land. Marinda and her brothers ice-skated on Boulder Creek, the banks of which I've walked so many times and which flooded in the 2013 "Hundred Year Flood," damaging much of the farmland in the county, including land that had been farmed by the Goulds and Platts.

Marinda and her immediate family left the homestead after only ten years to farm in the Black Hills of the Dakota Territory because they found farming to be too difficult in Colorado. From there they ventured to Nebraska. Theirs is a story like that of so many colonizing families. They would claim land, and if they did not strike gold or could not farm there, they would move on to the next place, where they hoped for better success.

The Native peoples who had been forcibly driven from what Marinda knew as Valmont and Boulder hadn't practiced the African and Euro-American agricultural methods on the land. The Ute, Southern Arapaho, and Cheyenne, who stewarded the land long before settler colonialists arrived, had followed migratory patterns, seasonally moving with the elk, deer, and bison herds between the eastern plains and the mountain ranges to the west of the Boulder foothills. The Southern Arapaho, in fact, were a farming people in Minnesota who had to abandon the practice by necessity when they were forced south to what became Colorado. They would winter in what is now known as Boulder County. Aside from game animals, they foraged wild herbs, plants, fruits, and roots for food and medicinal purposes. African and Euro-American agricultural practices were first used in the area when settlers like Marinda's uncle arrived during the Colorado Gold Rush, which began in 1858. The City of Boulder was founded in 1859, and the Colorado Territory was established in 1861.

Colonization of the area led to the forced removal and genocide of Native populations, most barbarically about 200 miles southeast at the Sand Creek Massacre of 1864, where a peaceful group of Arapaho and Cheyenne were encamped at the command of the American military. On November 29, 1864, American troops arrived at the encampment and brutally slaughtered 230 Cheyenne and Arapaho people. Those involved in the Massacre included men of Company D of the Third Colorado Cavalry—more than 100 Boulder County residents (46 from Boulder), who trained less than a mile from the Gould homestead on land that until 2023 bore a marker declaring the site "used during the Indian Uprising." In 2018 the City of Boulder acquired the land and is working "with representatives from Cheyenne and Arapaho Tribes of Oklahoma and the Northern Arapaho and the Northern Cheyenne who came to Boulder in late July [2018] to begin providing city Open Space and Mountain Parks staff input and feedback on the development of the [land management] plan" ("Fort Chambers").

Establishing a farm and homestead was a critical part of the colonizing mission. Although Native peoples had a rich heritage of agroecological practices (Peña), settler colonizers either failed to recognize these practices or in many instances deliberately destroyed plants and animals to starve out Native populations (LaDuke, *All Our Relations* and *Recovering*). "Claiming" land in Colorado, an idea of ownership foreign to Indigenous peoples who lived with the land in kinship, and "improving" that land claim through building a dwelling and cultivating the land, was required under the Homestead Act of 1862. In 1870, the constitution of the newly founded agricultural town of Longmont, Colorado, only a few miles from where the Goulds lived, read, "Agriculture is the basis of wealth, of power, of morality. It is the conservative element of all national and political and social growth; it steadies, preserves, purifies and elevates" (qtd. in Wolfenbarger 11). As Lisa King explains, for Native Nations, "[C]ulture and religion are in turn derived by the people from the land they inhabit; thus, the people, the culture, and the land take their meaning from each other" (19). When settler colonizers drove Native peoples from the land with which they lived, it was a destruction of culture on an embodied and spiritual level. It was a cultural genocide as well as a human genocide and an ecocide.

This is the place my ancestors claimed as their own. I would like to offer an acknowledgment: This land has never been ours to claim. All who live here are a part of it and its long, complicated history. In Boulder County's case, in discussions of local food, it is important to consider, as King urges, "how rhetoric is culturally constructed and the cultural impact particular rhetorics or discourses may have on the communities involved, particularly local Native communities" (32). I would like to acknowledge that when the majority of those who live here now discuss and define "local food" and "agricultural heritage," rarely are Native food cultures or native plants and animals included, though there are Native-led organizations and individuals here working to reintroduce and support Native foodways. Our language choices and definitions matter, as I will discuss throughout this book.

With humility I acknowledge the Native peoples who lived and still live here, as well as those forcibly relocated to Oklahoma, Montana, and elsewhere. I acknowledge the biodiversity that has been destroyed. I acknowledge those who died to protect this land, and the current Native Nations members who are working with Boulder County, the City of Boulder, and leading nonprofit organizations on issues from land use to renaming of sites to honesty in telling the histories of colonization here (Boulder County, "September"). For example, there are Native and BIPOC-led organizations such as First Nations Development Institute, Spirit of the Sun, Frontline Farming, Harvest of All First Nations, and others working toward food sovereignty and cultural sovereignty along the Front Range of Colorado. There are also members of the Cheyenne and Arapaho Tribes working with Right Relationship Boulder and the City of Boulder to determine how to use the site of Company D's training for the Sand Creek Massacre as a place of memorial and education. A first step in working toward accountability is to acknowledge that my ancestors' dream to farm and establish roots in what is now Boulder County, and that my own ability to live and work here now, are dependent on the uprooting of sovereign Native peoples. I must acknowledge the ongoing legacy of this violence, as I will explore in the pages of this book.

**

This book offers what it might look like for a person in a university or college to pursue research, teaching, and community-based work that align with deep connection to and exploration of place and communities, past and present. While some academics may find it desirable or necessary to change institutions or careers for any number of reasons, moving from place to place can be profoundly disruptive for family, for community building, and for mental health and happiness. Even knowing that long-term locational stability is not always possible or desired in the academic world, especially for graduate students, we can find transformative power in connecting deeply with the place where we find ourselves in the present moment.

Certainly, we do not need ancestral roots in a place for us to lean into this challenge, though in my case the ancestral connection to farming and settler colonization on my mother's side brought forth, in ways I am still learning to understand, my own calling into the study of local food and farming in Boulder County. Wherever we find ourselves, there are lessons we can learn as we attune to the particular place in which we live (Owens)—we can learn its natural patterns, plants, and animals, its histories, its original and current inhabitants, its cultural and social conversations, all of which make up parts of its ecology. We can be accountable to that place and work daily, as Kimmerer suggests, to unlearn and challenge ideologies of the colonizer—racism, individualism, hierarchy, scarcity, disconnection, power over—ideologies that regardless of our ancestry are infused into nearly every facet of daily life in the United States, into our universities and colleges (e.g., Kynard, "All I Need," "Oh No," and "Teaching"; Aja Martinez; Kannan et al.; Ruiz; Ore et al.), into our communities, into our own bodies and minds. And then, as we listen to place, our purposes as teachers, scholars, and members of our communities may reveal themselves in unexpected ways.

* * *

Ecology means "study of the house," from the Greek, *oikos*. *Oikos* refers not only to the physical place of dwelling but to family generation to generation as well. Marinda Platt—schoolteacher, farmer, settler, ancestor—lives on in me in inevitable ways as I tell a story of the land and its inhabitants' current forays into local food production and consumption. Inspired and held accountable by those who once lived here, by those who live here now, by those for whom I will one day be an ancestor, and for all those in the vast web that make up my ecology, I feel myself guided by *this place*. I open myself to its teachings and listen.

1

Food Rhetorics and a Crisis of Definitions

> *Let us shift our minds to regenerating the currencies that make our life work, and the currencies that come from nature as the original source. These are the currencies of food currencies, of water currencies, of bread, currencies of love, relationships, currencies of being cared for, and the capacity to care.*
> —VANDANA SHIVA, "DIVERSE EXPRESSIONS OF A LIVING EARTH"

The Ways in Which Companies and People Define Industrial Food

The current food crisis in the United States hinges directly on a crisis of definitions—our failure, as food scholar and journalist Michael Pollan would say, to attach "words to real things and precise concepts" ("Our Decrepit"). This failure to do so connects explicitly to an imbalance of power between people and corporations and highlights how Rhetoric and Writing scholars can offer a vital intervention in broader food studies and food justice conversations. In this chapter, I study the national discourses around various foods and food movements in order to explicitly connect food studies and rhetorical studies. I discuss definitional implications of language used by corporations and by people who challenge corporate language as I offer analysis of discourses about industrial food, followed by organic, local, and justice movements' uses of language. As a first step in *ecological community writing*, it is useful to understand the national discourses as parts of a food ecology as they will have a direct impact on any local discourses and will highlight the need for locally specific terms and definitions to be written by the people who live in the impacted communities.

FOOD CRISES WRIT LARGE

Big agriculture (also called "big ag") and big food corporations (also called "industrial food") seek to control the world's access to seeds, produce, meat, and water, influencing subsidy policies in the United States and lobbying for looser and fewer regulations. This has led to what some call the colonization of food, seeds, and water[1] (Shiva "Biopiracy"; Holt-Giménez), part of the legacy of colonialism during which Native populations' food sources have been stolen or destroyed as part of colonial violence and power grabs. As I have mentioned, this manifested early on in what is now the United States as "land claims," destruction of Native food supplies, and introduction of African and Euro-American agricultural methods. These practices continue today through, for example, market saturation of cheap products and genetically modified (GM) seeds that put Indigenous farmers from Mexico to India out of business unless they purchase those seeds, or that contaminate their heirloom varieties with GMO cross-pollination (LaDuke, *Recovering*; Shiva, "Biopiracy" and "Diverse"; Mihesuah and Hoover 5). Nationally, in the last thirty years, "more than 75% of USDA farm subsidy payments were received by fewer than 10% of farms," with larger commodity crop farms that grow GM varieties tending to receive a significantly larger share of the subsidy payments, marking a clear connection between big ag and government funding (Lewis 43).

As humans are increasingly dependent on the global industrial food system, our right to know what is in our food and how it is produced is eroding. Industrial food companies have weakened people's ability to access affordable, nourishing, safe food, and indeed, some argue, real food itself. Michael Pollan calls industrial, processed food "edible foodlike substances" or "notional" food ("Unhappy"; qtd. in *Food, Inc.*). Wynne Wright and Gerad Middendorf similarly state, "Utilizing efficient but unsustainable practices, the corporate food industry manufactures and markets 'artificial,' 'counterfeit,' 'anti-foods,' which are so homogenized and far removed from their 'natural' state that they are but a simulacrum of 'genuine food'" (qtd. in Shultz 225).

---

1 While international battles over water are beyond the scope of this book, for information on multinational companies' attempts to privatize the world's water supply, see Vandana Shiva's *Water Wars*, Charles Fishman's *The Big Thirst*, and Robert Glennon's *Unquenchable*.

This industrial food, constituting what Enrique Salmón calls the resulting "Big Gulp national culture" (*Eating* 78), has led to regular, lethal outbreaks of foodborne illnesses (Schlosser) and public health crises of obesity, diabetes, heart disease, and related illnesses, disproportionately in Black, Latinx, and Indigenous communities. Because these communities disproportionately only have access to food that is high in calories, and low in nutrient density, we see populations who are at once "stuffed and starved" (Patel), able to get food from neighborhood convenience and corner stores but not grocery stores (LaDuke, *Recovering*; Salmón, *Eating*). Karen Washington, Leah Penniman, and other food justice activists and farmers are reframing the language to describe this phenomenon. Rather than the common term "food deserts," which defines that lack of food in certain areas is a natural state, they call it "food apartheid," as the lack of access to healthy food is "a *human-created* system of segregation that relegates certain groups to food opulence and prevents others from accessing life-giving nourishment" (Washington; Penniman, *Farming* 4, emphasis original). The most heavily impacted are communities with large populations of Black, brown, and Indigenous people.

Lindsey Loberg, co-director of Boulder Food Rescue and one of my community partners in Boulder, discussed the importance of this distinction of terms and definitions in our interview together in 2020. Not only does Lindsey work on the front lines of food security and access work in Boulder, but they've also done public research on connections between how hunger manifests in Boulder and national, systemic issues. They explain, "An honest conversation about food access is inseparable from conversations about racism, white supremacy, ongoing colonialism, ableism, and economic systems that value financial profit over human wellbeing. These are root causes of food insecurity." Lindsey agrees with Karen Washington, who argues that the food system is not broken; rather, it works exactly as it is designed to "within larger systems of oppression that are intended to control and consolidate power" (Loberg). As they describe it in our interview, there are multiple food systems that operate simultaneously, the two most prominent of which they term the "choice system" and the "charitable system."

The choice system, as the term implies, is one where "people can have, for the most part, any food they want when they want it," Lindsey explains. The charitable system, which includes nonprofit food distribution, "is marked by a dramatic reduction in choice, power, and control." From years of community-based research and direct local experience with food rescue and distribution, Lindsey delineates what they observe daily: "When people use the charitable food system, it meets their most basic needs of survival around food, but there are still negative impacts on their physical health as well as their mental and emotional wellbeing. It creates an enormous amount of stress, emotional turmoil, and invisible labor for people. People do a tremendous amount of work to acquire food in ways that are frequently soul crushing and emotionally expensive. And they're still experiencing poor health outcomes, generally. So, the charitable food system is succeeding in routing people food but not in creating a reality in which people can thrive emotionally and physically. It's really difficult to move towards equity or gain political power if we're sick and exhausted and chasing around our basic needs" (Loberg). The connections that Lindsey, Washington, and other food justice activists make between the language people use to define food access and the systemic problems inherent in the industrial food system are critical for Rhetoric and Writing scholars to interrogate further.

Even definitions for justice-focused food access have expanded alongside movements in racial and disability justice. Roughly one-quarter of adults in the United States is living with a disability, and this can make access to food difficult. The USDA Economic Research Service reports year after year that "households that included an adult with disabilities reported higher food insecurity rates than households with no adults with disabilities." In 2022, it reported:

> For U.S. households that included an adult out of the labor force because of a disability, 28 percent were food insecure (low and very low food security). Among U.S. households with an adult age 18–64 who reported a disability but was not out of the labor force because of it, 24 percent were food insecure. In contrast, 7 percent of households with adults without disabilities were food insecure in 2021.

Where a definition of food access may for some mean availability or affordability of healthy food in a given area, for example, food justice disability advocates are raising awareness of the need to incorporate disability justice considerations into a robust definition of food access. This is an example of how expanding an ecology to incorporate previously marginalized and ignored voices offers *distributed definition building* that helps align a term with precise concepts.

The connection between health and food security has been demonstrated in several studies. For example, Alisha Coleman-Jensen and Mark Nord explain, "Disabilities present as a strong risk factor for food insecurity. This may be particularly detrimental for those with disabilities as food insecurity may exacerbate existing health limitations or health impairments." Even spaces that offer food access, from community gardens to food pantries to food assistance programs, are often created with an "able" body in mind. From paying attention to width between garden beds and shelf height in stores to signage with braille and raised images, disability justice advocates are working to fight ableism in food access and food justice conversations and realities. As Dana Ferrante explains, quoting scholar activist Natasha Simpson, "food justice organizations can avoid vilifying an illness, condition, or certain kinds of bodies. Instead, organizations can help '[link] communities' material conditions to structural inequities,' revealing ableism, rather than individual failure, to be at work when disabled people or those who are chronically ill are unable to access nourishing food or local community gardens." Tyler Martinez points out, "Individualism in food access upholds epistemic whiteness by reifying the illusion that individuals have power over systems of food access" (43). Disability justice in food justice conversations, much like the explicit connections between food justice and racial justice, has been all but ignored by populations not adversely impacted. The intersectional connections need to be centered more explicitly in future work in food literacies.

In addition to its human cost, the industrial food system has led to fossil fuel dependence for planting, fertilizing, harvesting, processing, packaging, and transporting food (Pollan, "Power Steer"; Brownlee, "Local Food and Farming" 2). It has also led to environmental destruction including topsoil depletion and contamination of our waterways from pesticides, fertilizers, and manure runoff, again causing harm to rural and Black, Latinx, and Indigenous communities disproportionately (W. Jackson, "Tackling"; Pollan, "Our Decrepit" and "Power Steer"; Alkon and Guthman). In an *ecological community writing* methodology, each of these elements and their impacts become part of the vast ecology that we both study and help to build through the connections our work can generate.

As I began writing this chapter in the fall of 2018, flooding in North Carolina due to Hurricane Florence had caused overflow of more than a dozen hog waste lagoons that spilled their contents into the waterways. Lagoons at concentrated animal feeding operations (CAFOs) are filled with dangerous contaminants: "CAFO waste isn't just manure, urine, and groundwater: It can contain birthing fluid, blood, hormones, chemicals like ammonia and heavy metals like copper (copper sulfate baths are used to clean the cows' [and pigs'] hooves); antibiotics put into their feed, and antibiotic resistant bacteria; pathogens like E. coli, cryptosporidium and salmonella" (Dobie 2016). Every year, there are stories of runoff from CAFOs entering waterways; Hurricane Florence is not the exception.

Other news from the same month featured Iowa corn farmers who were "terrified" about climate change's impacts on their agricultural yield (Cullen); Nebraska corn farmers who pled guilty to falsely selling their product as organic ("3 Nebraskans"), which led to contamination of the country's organic corn supply; three multinational food companies who sold their products worldwide despite positive tests for listeria and salmonella; and a recall of more than 12 million pounds of ground beef contaminated with salmonella. Unless we are affected directly, few people living in the United States will pay attention to stories like these that, every week, give glimpses into the realities of industrial food behind the veil of cheap abundance.

Despite the high costs to cheap food, relatively few Americans know about the contents of our food, let alone food production practices and their implications. Eileen Schell argues that the global industrial agriculture system "foster[s] an 'agricultural illiteracy' among the general public whereby the conditions under which our nation's food is grown, harvested, distributed, and marketed are made opaque and inaccessible" ("Rhetorics" 81). The general public is also unaware of how precarious and unresilient the industrial food system is (House, "Re-Framing" 3). And this ignorance is a deliberate condition that multinational seed and food companies work tirelessly to ensure.

THE INDUSTRIAL VEIL

When I think about industrial food, a scene from the film *The Matrix* comes to mind. It's the one in which the spiritual guide, Morpheus, asks the hero, Neo, if he wants to know what the matrix is. When Neo answers that he does, Morpheus says, "It is the world that has been pulled over your eyes to blind you from the truth." For me, this is the illusion of cheap and unlimited abundance on supermarket shelves and supersized portions at restaurants, the food simulacra.

The unquenchable desire for cheap abundance grows from an illusion deliberately perpetuated by multinational food corporations. The appearance of abundance is only possible due to the overproduction and waste built into the industrial food system. Because people might see large quantities of food in stores, Lindsay Loberg explains, "Studies show that [this visual] makes us buy more stuff, so it's more profitable. And because there are multiple steps in the food process that gets all that food to that point, waste is built into every one of those steps." Because grocery store consumers in the United States do not tend to want produce that is bruised or blemished or oddly shaped, food surplus, or what we may call "waste," is an essential part of the choice system.

Ironically, much of this food that is thrown out or donated to the charitable system is what people seem to love at farmers markets or from home gardens—the funny-shaped tomatoes with all of their splits, the twisted carrots—those "imperfections" show that the food isn't genetically modified or picked before it is ripe and sprayed with gases. But at the grocery store, these products don't make it onto the shelves. Supermarket abundance is about uniformity and "perfection." Over time and with careful marketing and crafting of definitions of what is and isn't desirable, companies have wielded tremendous power to change people's ideas of what good food looks like.

In the next several pages, I offer a brief overview of some of the problems that scientists and investigative journalists warn are central to industrial food and show how these problems are explicitly tied to issues of who controls food terms and definitions. In this way, I begin to build a case for Rhetoric and Writing practitioners' unique role as scholars and teachers of language and rhetoric. My method of *distributed definition building*, as I will discuss in subsequent chapters, brings the concepts we study to life in our communities.

Lifting the "world that has been pulled over [our] eyes" reveals what so many writers have called the high costs of cheap food—all of the externalities that don't show up at the register—and the unstable, unsustainable, and unhealthy nature of the industrial food system (Schlosser; Pollan, "No Bar Code," "Our Decrepit," and "Power Steer"; Kingsolver; Patel). When I say that the food system is unsustainable, I borrow Michael Pollan's argument about the modern food industry:

> To call a system unsustainable is not just to lodge an objection based on aesthetics, say, or fairness or some ideal of environmental rectitude. What it means is that *the practice or process can't go on indefinitely because it is destroying the very conditions on which it depends*. It means . . . that there are internal contradictions that sooner or later will lead to a breakdown. ("Our Decrepit," emphasis added)

Few people living in the United States realize that the way we eat is a major contributor to climate change and environmental degradation (*UN Climate Change*). As our food system has changed to feed people ever more "cheaply," the world's population, surging under this never-before-seen access to cheap calories, is growing exponentially at an unsustainable rate (Bartlett). Quite simply, we cannot sustain infinite growth on a finite planet with finite resources.

As the world's population continues to increase toward the estimated 9.7 billion people by 2050, according to the UN Intergovernmental Panel on Climate Change, "rapidly rising" demand for food is coupled with expected reduction of global food production by the end of the century due to climate change (17–18). According to the National Climate Assessment conducted in 2014 by 300 of the United States' most distinguished experts at the request of the US government, climate change is already decreasing agricultural production and capacity ("Climate Change 2014"). The 2018 *UN Climate Change Annual Report* opens with this sobering statement: "From shifting weather patterns *that threaten food production*, to rising sea levels that increase the risk of catastrophic flooding, the impacts of climate change are global in scope and unprecedented in scale" (emphasis added). Climate change brings the risk of significant shifts in terms of what and how much can be grown where.

No less important, and certainly related, is the belief shared by advocates of industrial food production, in the words of Richard Lobb of the National Chicken Council, in the "accomplishment" of "produc[ing] a lot of food on a small amount of land at a very affordable price" (*Food, Inc.*). Regenerative agriculture farmer Joel Salatin calls this the United States' belief in "fatter, faster, bigger, cheaper" (*Food, Inc.*). On the surface, this goal to produce cheap food appears logical and perhaps even ethical given the exploding population and decrease in land availability, as well as surging poverty and hunger rates. However, many scientists have been warning for decades of the substantial dangers of defining "accomplishment" in this way. The 2005 "Millennial Ecosystem Assessment" concludes that agriculture is the "largest threat to biodiversity and ecosystem function of any single human activity" (K. G. Cassman and Wood S., qtd. in W. Jackson, "50-Year" para. 7).

More recent warnings are no less dire. The opening to a June 2018 article in *Science*, based on the largest and most comprehensive study of its kind to date, states, "With current diets and production practices, feeding 7.6 billion people is degrading terrestrial and aquatic ecosystems, depleting water resources, and *driving climate change*" (Poore and Nemecek, emphasis added). The 2018 *UN Climate Change Annual Report* confirms that food production, specifically meat production, outscores even vehicle pollution in terms of its emissions and environmental footprint. As journalist Jonathan Foley explains, "[w]hen we think about threats to the environment, we tend to picture cars and smokestacks, not dinner. But the truth is, our need for food poses one of the biggest dangers to the planet." Indeed, the industry's belief in maximizing crop and animal production through large inputs of fossil fuels and chemicals and in defining the production in terms of innovation and accomplishment creates confusion and a cost far higher than most people in the United States are aware.

Fossil fuels are not the only natural resources being tapped by industrial agriculture. Soil is just as significant, though until recently few understood the depth of its importance. Big ag's approach to farming is based on the concept of monoculture—planting a single crop on a large plot of land. This approach necessitates the use of chemical fertilizers, pesticides, and herbicides, all of which strip the soil of nutrients. Much like the controversies over tar sands and other limited-output, extractive methods for attaining fossil fuels, Wes Jackson of the Land Institute explains, "The trouble is, the best soils on the best landscapes are already being farmed. Much of the future expansion of agriculture will be onto marginal lands where the risk of irreversible degradation under annual grain production is high. As these areas become degraded, expensive chemical, energy, and equipment inputs will become less effective and much less affordable" ("50-Year"). Jackson urges people, when they think of the scope of ecological collapse occurring today, to not only focus on climate change. He explains, "[C]limate change overshadows an ecological and cultural crisis of unequaled scale: soil erosion, loss of wild biodiversity, poisoned land and water, salinization, expanding dead zones, and the demise of rural communities" ("50-Year").

The industrial food system contributes to all of these crises, which is why the resurgence in regenerative farming that builds (regenerates) soil while sequestering carbon from the atmosphere is so vital.

Monocultures are not only detrimental to the soil but precarious too, as the cautionary tale of the 1845–1849 Irish Potato Famine revealed (Pollan, *Botany*). If we remain dependent on a very limited variety of genetically modified corn, soy, wheat, and other crops, as the Irish were dependent upon one type of potato, we risk global food shortages in the event of disease or crop failure. As a frame of reference, "ninety-two percent of Iowa's cultivated land is in just two crops—corn and soybeans" (Gayeton 29). According to the Center for Food Safety, "Currently, up to 92% of U.S. corn is genetically engineered (GE), as are 94% of soybeans and 94% of cotton (cottonseed oil is often used in food products). It has been estimated that upwards of 75% of processed foods on supermarket shelves—from soda to soup, crackers to condiments—contain genetically engineered ingredients." The repercussions of the failure of an entire industrial corn crop or soy crop could be catastrophic in terms of the global food supply. For these reasons and more, the work of heritage seed saving, often led by Black, brown, and Indigenous seed savers—for example, Rowen White of the Mohawk community of Akwesasne and Pat Gwin of the Cherokee Nation—is vital to diversification of the food system and restoration of cultural food traditions (e.g., Danovich; Kimmerer, *Braiding*; LaDuke, *All Our Relations* and *Recovering*; Penniman, *Farming*; R. White).

Americans' desire for cheap meat is also costly for a whole host of reasons, of which the general public are often unaware. In addition to vast use of resources, runoff from CAFOs, and soil depletion, there are human health consequences to the factory farm system of raising animals for food. According to the Union of Concerned Scientists, "at least 70 percent of the antibiotics used in the United States are fed to animals living on factory farms" (qtd. in Pollan, "Our Decrepit"), giving rise to antibiotic-resistant strains of bacteria such as *E. coli* O157:H7—a known public health threat (Pollan, "Power Steer"; Walsh). As Pollan warns, "[w]e inhabit the same microbial ecosystem as the animals we eat, and whatever happens to it also happens to us ... we are what we eat, it is often said, but of course that's only part of the story. We are what what we eat eats too" ("Power Steer"). In many ways, opposing ideologies are at play. As Pollan asserts in his discussion of factory-farmed meat, "[t]he economic logic behind the feedlot system is hard to refute. And yet so is the ecological logic behind a ruminant grazing on grass.... How cheap, really, is cheap feedlot beef? Not cheap at all, when you add in the invisible costs: of antibiotic resistance, environmental degradation, heart disease, E. coli poisoning, corn subsidies, imported oil and so on. All these are costs that grass-fed beef does not incur" ("Power Steer"). We pay one way or another; it is a matter of *where* along the chain we pay, and at least for financially secure citizens, that choice is partially based upon the paradigm through which we tend to view the world—an economic one or an ecological one. Food justice advocates have challenged this notion of choice, as I will discuss later in the chapter.

Why are so few people in the United States aware of these consequences of the industrial food system? Quite simply, multinational corporations work tirelessly to control the definitions and terms circulated to the public in order to ensure that the veil remains in place. Part of the strategy of a few of these companies has been fear based. As has been well documented, they have sought to bankrupt farmers they accuse of patent infringement; to delegitimize scientists who speak to potentially damaging effects of genetically modified organisms such as GM corn and soy; and to ensure that it is dangerous to speak out against industry (*Food, Inc.*; Shiva, "Biopiracy" and "Diverse"; Genoways).

Food libel laws, also known as food disparagement laws or veggie libel laws, in thirteen states have made it perilous for journalists and citizens to speak against the safety and contents of some foods as well as conditions under which food is produced, according to investigative reporting from sources such as NPR, *The New York Times*, and *Mother Jones*. In my home state of Colorado, it can be a felony offense to disparage the meat industry, despite First Amendment protection of free speech (Genoways 44). In many states, it is a crime to secretly videotape feedlots or to hold onto video without turning it over to authorities within forty-eight hours. This makes it very difficult to build chronic abuse cases (Bland). The result is an unstable food system that is increasingly illegible and undemocratic.

People's lack of food literacy relies upon a deliberate lack of or hiding of information. Anti-labeling campaigns, most recently focused on GMO labeling, lax definitions around current labeling requirements, and deliberately vague language in food marketing (think: "natural" or "farm fresh") have made it increasingly difficult for consumers to know what they are purchasing. In July 2015, the House Agriculture Committee passed H.R.

1599, which "not only den[ies] states the right to pass common sense GMO labeling laws, but also make[s] it illegal for local citizens to ban or even regulate GMOs in their states or counties" (Food Democracy Now!). As Eileen Schell explains, "we are losing our right to determine where our food comes from and how it is produced" (91). All of this leads to the questions of language's role in the confusion and of how rhetoricians can intervene in food studies conversations.

Definitions Matter

Lack of information does not only happen in these big ways mentioned so far but in everyday obfuscation, which many scholars have analyzed (Abisaid; Boerboom; Cooks; Adams; Miele and Evans). In his study of euphemisms in food industry marketing, such as "free-range" and "naturally-raised," Joseph L. Abisaid argues that "such labels are deliberately phrased to create a certain effect so as to minimize concern over the ethically questionable behavior of consuming animals and most importantly create a healthy conscience on the part of the consumer. Moreover, humane food labeling has the added benefit of making consumers feel good without asking too much of them" (145–46). Life on CAFOs is indefensible unless there is a lot of obfuscation and vagueness, Abisaid explains, and for this reason, industry has developed what he calls "the supermarket lexicon" (145–46). It is what Associated Press reporter Charlie Riedel calls "a way to distract people with feel-good messages" sometimes relying on "the blurring of marketing and advocacy." He cites, for example, Chipotle's unregulated term "responsibly raised," which "lets animals eat GMO feed." The capitalist imperative is to make desirable and palatable that which is at best unnecessary and at worst intolerable (Cooks). Playing on the familiar concept of greenwashing (Koch and Compton), Ellen Gorsevski offers the concept of meatwashing "as rhetorical discourses that obscure massive animal suffering required by globally burgeoning consumer demands for unsustainable amounts of meat" (203). Many years earlier, Pollan made a similar argument: "More than any other institution," he writes, "the American industrial animal farm offers a nightmarish glimpse at what Capitalism can look like in the absence of moral or regulatory constraint. Here, in these places, *life itself is redefined*—as protein production—and with it, suffering" ("Animal's Place," emphasis added).

This "redefinition" is not reserved for meat. Erin Trauth relays a 2008 court case against Gerber fruit snacks' false claim that their product contained the fruit featured on the front of the package. "The court found that 'reasonable consumers should [not] be expected to look beyond misleading representations on the front of the box to discover the truth from the ingredient list'" (*Williams v. Gerber* 2008, qtd. in Trauth 9). Similar claims can be made about all kinds of packaged food, whether meats or grains—companies *expect* that consumers will not look too closely and will be (as Morpheus suggests in *The Matrix*) contented by the world pulled over their eyes. Douglas Gayeton says, "[y]our local supermarket sells more stories than your local bookstore. And just like your bookstore, these stories—an artful mix of fact and fiction—are placed on shelves" (81). These stories create the narrative of cheap abundance and veil the realities of what we are eating and the systems in place to make those stories palatable.

Stanley Deetz proposes that "people's non-work lives have been systematically colonized by corporations and corporate values." Deetz also argues that this has happened through the colonization of language that has become "systematically distorted by corporations with the power to control communication in a capitalist society" (qtd. in Koch and Compton 231). The Bayer Group's (in 2018 Bayer and Monsanto merged and rebranded) website offers an example of this corporate distortion of definitions. By using language that environmentalists and organic farmers use, they claim the language for themselves and obscure the meaning of the words. Their website says, for example, "There has never been a more important time for innovation in agriculture. Our world faces enormous challenges including a changing climate, limited natural resources, and a growing population. And we believe agriculture is part of the solution." On their "Food Security" page, they state,

> Our agricultural initiatives and products are aimed at making a contribution to one of society's greatest challenges—food security. Every second, the world's population increases by another three people. By the year 2050, there will be almost 10 billion humans on this planet. We want to help ensure a safe supply of food, now and in the future. We consider the achievement of this objective to be inextricably linked with sustainable agriculture.

Their language sounds like that of the climate and agriculture scientists I quoted earlier, but by co-opting the very real concerns for their own agenda they confuse consumers into believing without questioning. In the process, they perpetuate legacies of violence rooted in colonial frameworks. Their control of language and definitions in discussions of sustainable and innovative agriculture is central to the problem.

Let's take as an example how they are defining "sustainable" food production. Their claim that the only way to feed the world is through their version of "sustainable" agriculture is particularly problematic when we consider that the industrial food system uses 70 percent of the world's agricultural resources to provide only 30 percent of the world's food. In addition, 80–90 percent of the corn grown in the US is used to produce ethanol and animal feed, not to feed humans (UN Report). Of the 10–20 percent used for human food, at least 30 percent of that is used to make high-fructose corn syrup. Studies from the United Nations, Rodale Institute, and *Agriculture and Human Values* "demonstrate that crop yields using organic and sustainable methods are competitive with yields derived from [industrial] technologies" (Katz 375). I do not discount the potential of genetic modification, for example, to produce drought-tolerant crops or crops that can withstand higher temperatures. But the claim that GM corn and soy seeds exist primarily to "feed the world" is simply not true.

Bayer's promotion of the herbicides and pesticides used on GM crops that contain substances (such as Roundup's main ingredient, glyphosate, which the World Health Organization has deemed a "probable carcinogen") does not support their claims of valuing transparency, the health of consumers and farmers, and sustainability for the planet. It is also true that because of redlining and systemic racism in city and neighborhood planning, people of color are significantly more likely to live "in neighborhoods dominated by toxic industries and diesel emissions, or in rural areas burdened by the pesticides and dust that result from agribusiness" (Alkon and Agyeman 7). Through domination of the market, bankrupting small farmers who refuse to grow GM crops, outlawing seed saving, making independent research on their seeds extremely difficult, and other practices, they hardly live up to the values they promote on their website. In their effort to disseminate their information, they also aim their messaging at children, from activity/coloring books to online games, which tout the benefits of GMOs. Much research has shown that while yields have increased due to these intensive kinds of agriculture, so has harm to farmers, the environment, communities, and food security (Shiva, "Biopiracy," "Diverse," and *Who Really*; Magdoff et al.; Born and Purcell). None of this is sustainable.

Companies benefit from abusive systems of labor framed through language of "efficiency" and "output" and "cheapness." Human beings become like machines to be pushed to the limits of their capacities, and they are interchangeable and disposable. Many deaths and injuries have resulted from extremely dangerous and fast-paced meatpacking plant conditions (Schlosser); slavery and horrific farmworker abuses in the tomato fields of Florida (Estabrook); and deadly contamination of air and water with manure, pesticides, and other toxins in communities near production facilities (Alkon and Guthman; LaDuke, *All Our Relations*). The legacy of racist, ableist, and colonial practices undergirds the illusion of "cheap" food. Continued acts of food-related violence against Black, brown, and Indigenous peoples are current manifestations of centuries of white control of food supplies tied to subjugation of human beings. The current industrial food system is built upon injustices against farmers, farm and factory workers, animals, and whole communities, shrouded in the language of innovation, efficiency, sustainability, and cheap abundance.

Powerful multinational food and agriculture companies have claimed the role of literacy sponsor. In *Literacy in American Lives*, Deborah Brandt defines literacy sponsors as "any agents, local or distant, concrete or abstract, who enable, support, teach, or model, *as well as recruit, regulate, suppress, or withhold literacy—and gain advantage by it in some way*" (emphasis added). When big ag and big food label their products in order to confuse; when they fight to outlaw labeling that would inform; when they suppress literacy about the practices that get food to our tables; when they knowingly market as "healthy" the food that makes us sick and promote their own definition of what responsible and sustainable food production entails, they wield their power to sponsor their version of food literacy. It is up to all of us in our multiple roles—community members, rhetoricians, teachers, and students—to counter what Elspeth Stuckey calls the "violence of literacy" and to reclaim our own right to understand what is in our food, how it is grown, and how it connects to systems of exploitation and disease. We can begin to do this through a study of language and definitional control.

Defining "Organic"

Many of the concepts of organic agriculture were originally introduced by enslaved peoples who brought traditions of crop rotation, composting, and other practices from Africa, and were circulated more widely by Black farmers like George Washington Carver, Booker T. Whatley, and others (Penniman, "Why Food"). Indigenous populations also had sophisticated methods of "cultivation of indigenous plant medicines and foods" deliberately wiped out by what Winona LaDuke calls "a scorched earth plan to destroy all crops and livestock" (*Recovering* 202, 196). While sustainable agriculture was practiced around the world for thousands of years, the term "organic agriculture" was not coined until 1940 as the use of chemicals and pesticides was beginning to gain traction in agriculture. During World War II, chemicals designed as nerve agents were found to be effective in killing pests on crops, and industrial agriculture, which would rely heavily on these chemicals, took off in earnest. Vandana Shiva points to the ongoing legacy of war used in the language and naming of pesticides:

> Monocultures and monopolies symbolize patriarchal agriculture. The war mentality underlying military-industrial agriculture is evident from the names given to the herbicides destroying the economic basis of the survival of the poorest women in the rural areas of the Third World. Roundup, Machete, and Lasso from Monsanto. Pentagon, Prowl, Scepter, Squadron, Cadre, and Avenge from American Home Products, which has merged with Monsanto. The language is of war, not sustainability. (*Staying Alive*, "Introduction")

As David Nowacek and Rebecca Nowacek explain in their study of the discursive elements of the organic movement, the real growth in "interest in organic foods was deeply immersed in the anti-military-industrial complex counterculture of the late 1960s and early 1970s. Many organic consumers and farmers entered the organic foods system as a way to resist conventional agriculture" (414). "Organic" became a word of the counterculture and was originally synonymous with small-scale, family-owned, sustainable farming. As such, it seemed an answer to its industrial counterpart with the military legacy. The Rodale Press offered the first organic certification in the 1970s. In 1974, farmers took over the certification, becoming California Certified Organic, and other regional certification programs began to crop up around the country.

It did not take long, however, for big ag to get into what has become known as "big organic." By the mid-1980s, "the forces of market competition and of the lure of profit soon attracted [companies with] motives that differed from and were sometimes antithetical to the motives of earlier participants in the organic foods system" (Nowacek and Nowacek 412). In 1990, a section of the Farm Bill–Organic Food Productions Act, Title 21, charged the USDA with defining and regulating the organic food market and the standards for organic production. There ensued a decade of debate, and in 2002, they issued their set of criteria for the USDA-certified organic label.

By all accounts, organic's growth has been remarkable—from about $17 billion in sales in 2006 to upwards of $49 billion in 2017. As the Organic Trade Association (OTA) puts it, "[c]onsumer demand for organic has grown by double-digits nearly every year since the 1990s." This has happened so quickly that the demand now exceeds the supply. According to the OTA's 2018 Industry Report, "[o]rganic sales in the U.S. totaled a new record of $49.4 billion in 2017, up 6.4 percent from the previous year and reflecting new sales of nearly $3.5 billion. . . . [I]t was well above that of the overall food market, which nudged up 1.1 percent." In 2023, the Organic Trade Association reported that the US organic food market was $63.8 billion in sales that year.

When the market looks promising, big ag and big food are not far behind. In their move to enter the organics market, they have continued to fight for more lax definitions. They are "producers who are 'organic' only by virtue of minimal compliance with standards that no longer embody the original visions of organic agriculture" (Nowacek and Nowacek 403). In fact, big organic has few of the environmental and societal benefits that small-scale, sustainable organic farming does. Like conventional industrial agriculture, "big organic" relies on monoculture and large quantities of fossil fuels in production, packaging, and transporting. Data from several studies indicate that the larger the grower, the more minimally they tend to comply with standards (Guthman; Rosset and Altieri). Nowacek and Nowacek posit that "the market-driven erosion of the environmental benefits of organic agriculture has opened a growing gap between the benefits consumers believe they are promoting and the benefits they are actually promoting"

(403–404). Along these lines, a 2012 UK study showed "a wide discrepancy between participants' perceptions and evaluation of sustainable food labels but reported that participants found many labels to be trustworthy even if they had no idea what the label actually meant" (Koch and Compton 231).

Many who pay attention to organic's growth worry that both the word's definition and the industry have been co-opted by the very systems they wished to avoid and that it no longer signifies a transgressive or transparent movement. In addition, while some growers use organic practices that consumers may associate with what they define "organic" to be, those growers cannot afford certification or have opted out for philosophical or political reasons. It may be, then, as Michael Pollan warned, that the word is no longer associated with "real things and precise concepts." Some growers call themselves "beyond organic," as they identify with the organic movement rather than the organics industry (Gayeton) or use principles of regenerative agriculture, which integrates plants, animals, soil health, and biodiversity (127). Michael Sligh, owner of Foxglove Farm in British Columbia and founding chair of the National Organics Standard Board, explains the shifting definitions associated with the term "organic": "'Organic' was ... to identify a broad set of social, ecological, and spiritual principles about our farms and how we produced our food for our communities. Now the USDA has given the word 'organic' a legal definition, in essence, taking ownership of the word, and limiting its use to a narrow set of rules and regulations designed to support a distribution and marketing system. For some of us the word no longer addresses the deeper issues that were at the heart of the movement" (qtd. in Gayeton 86). As Eliot Coleman explains in his essay "Beyond Organic," organic certification "is tailored to meet the marketing needs of organizations that have no connection to the agricultural integrity organic once represented" (113). With clear problems of definition, what *does* organic mean?

Despite all of these problems related to how organic is and could be defined, it is critical to note how ecologically important organic practices in agriculture are, however flawed "big organic" may be. For example, if a product is certified USDA Organic, it cannot contain genetically modified organisms (GMOs), which is important to many consumers. GMO seed companies are currently fighting to enter the organics market and change the official government definition for what "organic" certification means. Certified organic products also will not contain the petrochemical-based pesticides that conventional products may hold. In some conventional fruits and vegetables such as berries, apples, and celery, the pesticides seep through to the center of the food, making them some of the most pesticide-laden foods on the market. Experts argue that these foods and other produce, particularly foods included in the Environmental Working Group's annual "dirty dozen" publication, are especially good to buy in their organic varieties when possible, as chronic pesticide ingestion may contribute to health problems, including cancer.

Undeniably, organic food is "a market that allows consumers to indicate that they are willing to pay more for food that is produced under environmental (and perhaps social) conditions that they find more palatable. This is a significant corrective to the market" (Nowacek and Nowacek 405). Indeed, for those with the means to buy organic, the purchase of organic products is a rhetorical act premised upon a set of beliefs often associated with environmental, health, and community-centered concerns. Abisaid notes, "The last few decades have witnessed a type of grassroots movement of food 'conscientious consumers' that have demanded greater transparency in the food production process" (141). It is a rejection of one set of beliefs and practices and an embrace of another set.

But this "vote with your fork" rhetoric of individual responsibility only holds in populations with the financial means to participate. It excludes the millions of people left out of this scenario because we as a country fail to make organic food affordable and accessible; heavily subsidize corn, soy, and other commodity crops but not vegetables; and do not provide a living wage to all working people. How do we work to make corrections to this vast, fundamentally flawed system? Alison Alkon and Julian Agyeman challenge the rhetoric of individual corrections to the market "in favor of a more environmentally sustainable food system to imagine that all communities, regardless of race or income, can have both increased access to healthy food and the power to influence a food system that prioritizes environmental and human needs over agribusiness profits" (6). Food choices are not always individual choices—they reflect structural realities, and there are clear connections between hunger, systemic racism, ableism, inequity, and language. A problem with the industrial food system is that "privileged people get higher quality food while poor people get the cheapest food or food that is left over . . . from the primary market and is usually inferior in quality" (McEntee 249). I will return to the question of systemic inequities later in the chapter.

Because of some of the concerns about big organic that make it difficult for consumers to know the real practices behind their food production; because it is hard for the USDA to measure some of the metrics that consumers say that they care about; and because multinational companies sometimes source their organic products from other countries with more lax definitions of "organic" in order to market their product more cheaply, many people have turned toward a method of sourcing that seems more transparent—buying and growing local food. As Smith and Denton explain, "[w]hen a movement has succeeded in becoming part of the establishment, 'the movement-turned-institution will face a new generation of reformers and revolutionaries who become dissatisfied with the new order'" (qtd. in Hahn and Bruner 51). For many food advocates and activists, "local" became the word of this new generation.

I don't want to overstate the rejection of organic for local. Organic farming is a vital part of the food equation, not to mention that it is virtually impossible for people in many parts of the country to live on a purely local diet. In Boulder County, where I live, eating locally in the off-season means mostly foraged and preserved foods, root vegetables or produce grown indoors, and hunted or locally farmed meat. Many people in other parts of the country do not have even that kind of variety available. But with local purchasing, the *ideal* (while not always possible) is that we can, as the slogan says, "know [our] farmer." We can visit the farm or urban growing space or neighborhood gardens. We can ask about growing practices. We can see for ourselves whether the practices align with our personal values and make purchasing decisions accordingly. We can learn what is in season when and incorporate seasonal food into our diets. We can learn how to save seeds and grow some of our own food, even if on a small balcony, or grow in community spaces based on cultural and dietary preferences. At least some of these options are often available even in dense cities or rural towns without nearby grocery stores.

Defining the Local Food Movement(s)

Richard Heinberg, senior fellow at the Post Carbon Institute, concludes in his starkly titled "What Will We Eat as the Oil Runs Out?": "The transition to a fossil-fuel-free food system does not constitute a distant utopian proposal. It is an unavoidable, immediate, and immense challenge that will call for unprecedented levels of creativity at all levels of society" (13). Local food movements draw on rhetorics of collective responsibility to take action against multinational corporate giants, and environmental degradation; at their best, they recognize systemic racism's connection to the industrial food system. They can act to anticipate shocks to the food system such as agricultural shifts due to climate change or impeded access due to fuel depletion, natural disasters, or pandemics. Having emerged as an antiestablishment antidote to big organic, these are often "know your food" and "know your farmer" campaigns about reclaiming food democracy and about supporting ethical and regenerative practices (Alkon and Guthman; Penniman, *Farming*; M. White). As a possible answer to what many view as an unsustainable and unhealthy food system, local food movements have grown exponentially since the mid-2000s.

The creator of the word "locavore," defined as someone who eats a diet heavy in locally grown foods, is Local Foods Wheel co-creator Jessica Prentice. In 2005, she and two other San Francisco Bay Area women, Dede Sampson and Sage Van Wing, challenged Bay Area residents to eat only foods grown or harvested within a 100-mile radius of San Francisco for the entire month of August. Prentice coined the term "locavore" for this challenge. In one of the earliest online publications about local food, Jennifer Maiser, editor of the Eat Local Challenge website, offers the first definition: "[L]ocavores are people who pay attention to where their food comes from and commit to eating local foods as much as possible . . . The first step to being a locavore is to determine what local means to you. This is an individual decision . . ." Following Prentice, she suggests a 100-mile radius: "The local food movement embodies citizens' understanding (however articulated) that the human food system is in crisis and that they want to re-gain control over what they eat. It is a movement aiming toward food democracy." In early writings about the movement, "local food" serves as a material, rhetorical embodiment of an answer to a global food system that deliberately obfuscates the processes and contents of industrial food and that encourages consumer disengagement and misinformation. The concept caught on quickly, and in 2007 the *New Oxford American Dictionary* named "locavore" their "Word of the Year" (Katz 376). In the same year, *Time Magazine* featured a cover story: "Forget Organic, Eat Local."

Soon, the government got involved in definition building. According to the 2008 Farm Act, a product can be marketed as locally or regionally produced if its end-point purchase is within 400 miles from its origin, or within state boundaries (S. Martinez et al., qtd. in Brain 1). While the USDA suggests a 400-mile radius for their definition of "local," there is no official "local" designation, certification, or set of criteria. In 2009 the United States Department of Agriculture established the "Know Your Farmer, Know Your Food" campaign to promote production and consumption of local food.

In many ways, local food movements offer hope for localized resilience in the face of what some anticipate as potential global food shortages. As Barbara Kingsolver points out, even an organic, vegan industrial diet can have a serious environmental footprint: "Bananas that cost a rainforest, refrigerator-trucked soymilk, and prewashed spinach shipped two thousand miles in plastic containers do not seem cruelty-free" (225). Eileen Schell offers a similar argument: "In an age of depleted fossil fuel resources and global warming, the logic of 'global food' has reached a critical juncture and has caused many rural advocacy organizations and environmental groups to argue for a return to local and regional systems of food production" (91). Local food advocates suggest that to eat locally sourced food is to shift from passive to active consumer. Figuring out how to afford eating healthy foods that align with ethical considerations is something that must be addressed, and I believe that it is most effective to do so at a local level. Scientists such as Wes Jackson acknowledge, "[E]ach region has its own problems and opportunities. We must acknowledge that all successful corrections will be local" (Jackson, "The Necessity").

The compelling arguments for local food have, perhaps inevitably, led to growth in demand. In 2017 *Business Insider* urged investors to pay attention to upward trends in local food (Hesterman and Horan). In her article about the future of local, "The Local Food Movement 15 years In," Marylin Nestle explains, "The USDA is mainly devoted to promoting industrial agriculture but has had to pay attention (if a bit grudgingly) to the growth of local and regional food systems. It reports that about 8 percent of U.S. farms market foods on the local level, mostly directly to consumers through farmers' markets and harvest subscription (CSA) arrangements. It estimates local food sales at more than $6 billion a year. This is a tiny fraction of U.S. food sales but growing all the time" (Nestle).

Local food has become so powerful a concept in the cultural zeitgeist that now, much as they did with organic, even multinational corporations are adopting local rhetorics, strategically deploying the adjective in marketing campaigns. McDonald's launched an advertising campaign that acquaints consumers with growers, fishermen, and ranchers who supply to the food Goliath. Their website states: "Trace your food back to THE SOURCE" (capitals in original). "We care about where our food comes from. That's why we use responsibly grown ingredients from the McDonald's community of farmers. Get to know a few of our suppliers who make delicious food you can feel good about." In one ad, we learn: "From the dock at Dutch Harbor, Kenny Longaker sets out on his boat, the Defender, to catch the wild Alaskan Pollock used in our Filet-O-Fish® sandwich." In another ad, we are told that "[t]hrough hard work and dedication, Frank was able to achieve his dream of becoming a potato farmer and owning his own land in Warden, Washington. Today, the farm produces a bountiful crop of Russet Burbank—one of the best potatoes for McDonald's World Famous Fries®."

Similarly, Walmart's "neighborhood markets" have sprung up around the country, and Walmart launched a program called "Heritage Agriculture," which "will encourage farms within a day's drive of one of its warehouses to grow crops that now take days to arrive in trucks from states like Florida or California. In many cases the crops once flourished in the places where Walmart is encouraging their revival but vanished because of Big Agriculture competition" (Kummer 40). In some cases, Walmart even works with nonprofits and universities to solve logistical problems with supporting local ag: "how much a relatively small farmer can grow and how reliably, given short growing seasons; how to charge a competitive price when the farmer's expenses are so much higher than those of industrial farms; and how to get produce from farm to warehouse" (40). I call this kind of campaign "Big Local."

Some food justice advocates would ask that before we praise this work too enthusiastically, we consider, "besides the location and access issue, the nature and impact of the operation itself, whether in relation to food source and supply chain, working conditions, or other environmental and land use issues" (Gottlieb and Joshi 156). While Walmart does have local food offerings, we should also acknowledge that nearly 10 percent of all Chinese exports end up in a Walmart, resulting in a loss of hundreds of thousands of US jobs (157). In addition, while their 2008 "Buy Local" promotion strategy did offer cheaper local products, it was at the expense of local food networks that could not compete (160).

This is where we bump into some of the arguments against blanket definitions of local food. One of the most convincing articles about concerns with local food rhetorics comes from urban planning scholars Branden Born and Mark Purcell, who warn against falling into the "local trap." They explain that consumers often assume that "a local-scale food system will be inherently more socially just than a national-scale or global-scale food system . . . the local trap is the assumption that local is inherently good. . . . Local-scale food systems are equally likely to be just or unjust, sustainable or unsustainable, secure or insecure" (195). Importantly, they explain, "the local trap conflates the scale of a food system with desired outcome. . . . [I]t confuses ends with means, or goals with strategies. It treats localization as an end in itself rather than as a means to an end, such as justice, sustainability, and so on" (196). Their research turned up ways in which local food systems resulted in "environmental degradation [and] exacerbated inequality" (196). Their concern is "an essentialized view of scale that sees global as hegemonic and oppressive and the local as radical and subversive" (200). They raise concerns about the assumed environmental sustainability of local; for example, "Consider a buy-local campaign in Arizona. Any ecological benefit from using less fuel for transport clearly would be outweighed by the need for massive water inputs" (200). They urge against the uncritical conflation. As with Pollan's argument about problems with usage of "natural" in defining food, "Implicit here is the idea that nature is a repository of abiding moral and ethical values—and that we can say with confidence exactly what those values are"—so can we say of "local" ("Why 'Natural'").

Additionally, as I shared through the story of my own family lineage, an assumption that local food is historically the native food of the area may overlook the place's colonial past and perpetuate the legacy of erasure of Indigenous food cultures (which, to address Born and Purcell's concern, did survive in what is now Arizona, in fact, through dry land farming and other ancestral techniques), as well as the violent histories that led to current local food definitions. But what the food behemoths like McDonald's and Walmart cannot appropriate through their marketing gimmicks, and what critics may misunderstand, is the "real, primal, fundamental connection [to our food] that [Americans have] been missing, and particularly in this virtual world. [Local food] is a tremendous counterbalance," according to Boulder County Farmers Markets executive director Brian Coppom ("Interview with Boulder County Farmers Markets" 82). If that need for connection can be coupled with education about history of place and systems supported through our purchases, local food as a rhetorical concept as well as a material thing has potential to open conversations about its connections to critical issues, ethics, and visions for food systems that align with accessibility and justice rather than financial ability. As I will discuss beginning with the next chapter, rhetoricians are well suited to intervene in these conversations through *distributed definition building* in local communities.

Logic- and Values-Based Rhetorics

As humans we all once understood ourselves, and many people across the world still understand themselves, through our relationship to the natural world. In modern times, through cultural, spiritual, and physical genocide, broadly speaking as a national culture, the United States has elevated innovation, convenience, individualism, and wealth accumulation, collectively considering the movement away from relationship with land as a mark of progress.

But a countershift is occurring in multiple arenas and on multiple levels of consciousness, representing both a political and cultural stance as well as a deep, spiritual need for connection and reclamation. As Prentice, the creator of the word "locavore," explains, "the beauty of the local food system is that it brings you back into a relationship with the source of your food, with the land, the animals, the plants, the farmers, and with each other" (30). As a society writ large, modern people living in the United States have forgotten that we are part of the earth, not separate from it.

On the other hand, many Indigenous Nations believe humans are in reciprocal relation or kinship with other beings, whose lives are sacred (Kimmerer, *Braiding*; M. Powell; Riley Mukavetz and Powell, "Becoming" and "Making"). In this worldview, human survival is directly correlated to the bounty of the land. As Robin Wall Kimmerer delineates in her explanation of the Honorable Harvest, we must "give back, in reciprocity, for what we have been given" (*Braiding*, 190). Kimmerer teaches, "The market economy story has spread like wildfire, with uneven results for human well-being and devastation for the natural world. But it is a story we have told ourselves and we are free to tell another, to reclaim the old one" (31). People living in the United States can turn away from the deeply ingrained rhetorics of efficiency, growth, hoarding of resources, perfectionism, and disconnection. Other models of existence can be embraced.

Though there are certainly emotions involved in people's food choices, some of the rhetoric about shifting these decisions is more focused on providing knowledge. By the early 2000s, Michael Pollan became a thought leader in this kind of rhetorical vein that some food justice scholars have criticized as a well-to-do, white-oriented framework. I appreciate Pollan tremendously for how his writings have shaped my consciousness, and I teach several of his works in my Rhetoric and Writing courses while also appreciating and teaching the important criticisms of his earlier perspectives, which have inevitably evolved in the last twenty years since he published *The Omnivore's Dilemma*. I share some of his early ideas here to delve into the "increase-people's-knowledge" arguments that became prevalent in many food-movement circles. Pollan writes, "[T]he omnivore's dilemma isn't simply a question of what foods are 'good to eat,' but also what foods are 'good to think'" (289). Pollan describes "the renaissance of local food systems, and the values they support, values that go far beyond the ones a food buyer supports when he or she buys organic in the supermarket" ("No Bar Code"). He frames the growth of local food systems as based on the value of "integrity" as opposed to the deception inherent in industrial systems. "Much of our food system," he explains, "depends on our *not knowing* much about it, beyond the price disclosed by the checkout scanner; cheapness and *ignorance* are mutually reinforcing" ("No Bar Code," emphasis added). I agree with him. The food system operates as it does through deliberate obfuscation and distortion of what words really mean.

Criticism of his writings comes from the leap Pollan then makes: "The more knowledge people have about the way their food is produced, the more likely it is that their values—and not just 'value'—will inform their purchasing decisions" ("No Bar Code"). In other words, Pollan argued that individuals will learn and then act accordingly. Many of Pollan's books and articles in *The New York Times* and *Mother Jones* appeal to an educated, progressive, financially secure demographic, which, he assumes, will already care about environmental issues to some degree even if they have not yet made the connections that he suggests between food consumption and environmental degradation. He explains, "[T]he decision to eat locally is an act of land conservation as well, one that is probably a lot more effective (and sustainable) than writing checks to environmental organizations" ("No Bar Code"). He wants his readers to understand that "Local food, as opposed to organic, implies a new economy as well as a new agriculture—new social and economic relationships as well as new ecological ones" ("No Bar Code"). For Pollan, eating "right" is a natural choice when we're faced with knowledge about the far-reaching consequences of our food decisions. Pollan distinguishes between "an industrial eater" and what he calls

> a new kind of eater—one who regards finding, preparing, and preserving food as one of the pleasures of life rather than a chore. One whose sense of taste has ruined him for a Big Mac, and whose sense of place has ruined him for shopping for groceries at a Wal-Mart. This is the consumer who understands—or remembers—that, in Wendell Berry's memorable phrase, "eating is an agricultural act." He might have added it's a political act as well. ("No Bar Code")

While he is certainly right for some people, Pollan and several other food movement advocates with similar ideas have faced criticism for the prevalent rhetoric that assumes that knowledge is the key determinant for how people eat.

Critics, whose work I discuss shortly, have argued that there are structural inequities that make this choice less than clear cut. In addition, while Pollan uses the words "renaissance" and "remember," he also posits the newness of the agriculture, eating, economy, and relationships he proposes. While the shift he suggests will be "new" for many of the people in the United States he is writing to, and I think that is what he's getting at, the concepts are as old as human agriculture has existed (dating to about 10,000 BCE) and are still practiced by cultures around the world, as Kimmerer acknowledges when she urges us to "reclaim the old."

The knowledge-driven vein of discourse that posits that the more accurate the language we use to discuss food issues, the more knowledgeable people's food choices will become led Gayeton to create what is now well known in many local food circles as "The Lexicon." Gayeton argues that we need a local food lexicon because "if people don't understand the meaning (and implication) of terms . . . how can they live more sustainably?" (10). He worked with 200 thought leaders, "architects of a new vocabulary," "to help 'take back' the meanings of these important ideas" (10). He writes, "[B]ut scare tactics won't bring about a shift in consciousness. People run from bad news, not toward it. The answer is to build consensus on a foundation of innovative ideas and solutions. Find out what works, then let solutions spread" (18). Gayeton's lexicon attempts to create an alternative discourse that allows people to understand one another without corporate intervention or co-option.

Several scholars and community organizers (e.g., Alkon and Agyeman; Garth and Reese; Penniman) have pointed out that this lexicon or discourse must expand its scope and the definitions it offers to include an antiracist, justice orientation and must attack systemic injustices that make it so difficult to eat healthily. The local food movement should not be reserved, they argue, for only those who can afford it, and it should not reify whiteness or ableism or colonial power. For some eaters, Gayeton and Pollan are right: They learn, and then they make purchasing decisions accordingly. But many people who have developed their food literacy cannot act for financial reasons or due to lack of access.

Many food justice scholars have critiqued "the unbearable whiteness of alternative food" (Guthman 263)—and the corresponding rhetorics of some well-to-do (mostly white) writers and activists that if people only knew, they'd make different choices. Guthman asks, "Who is the speaker? How do we identify those who do not know the source of their food? What would they do if they only knew? Do they not know now?" (263). That people should be willing to pay the real cost of food is a privileged, exclusionary discourse that ignores those who cannot afford to shop their values or who have no access to food that aligns with their values. In contrast to Pollan, Alkon and Agyeman argue that "market-based solutions are inherently undemocratic, as participation in them requires money" (341). Robert Gottlieb and Anupama Joshi's 2010 book, one of the first full-length studies to bring justice into the discussion on food, sets out "to identify a language and a set of meanings, told through stories as well as analysis," and part of their work is to analyze "the shift toward a new language" (7). Again, definitions (and what they include or do not include) matter. I am interested in the connection between community food literacies and the desire for a food language that embodies in its definitions an understanding of justice issues.

The shift in language is an essential component that both reflects and propels developing literacies. Lynne Bloom posits that "[n]ative eaters, like native speakers, learn from birth the cultural grammar of the language and employ it automatically in a host of contexts determined, in part, by when they live . . . , where they live . . . , how much they can spend . . . , what is appropriate to the occasion" (346). She describes a learned lexicon, morphology, and syntax of food. It is not just that there are learned discourses around food but that food is itself a discourse, quoting from Italian food historian Massimo Montinari, who posits a grammar of food (347). Eileen Schell agrees that to combat "the logic of 'global food,'?" we must "seek alternative discourses and advocacy rhetorics that offer food consumers and farmers sustainable alternatives" (92).

What is the alternate discourse needed? According to the Community Food Security Coalition, community food security "is a condition in which all community residents obtain a safe, culturally appropriate, nutritionally adequate diet through a sustainable food system that maximizes community self-reliance and social justice" (qtd. in Katz 373). The food movement cannot be characterized as simply a hipster or white movement, but rather, in its best iterations and in a more capacious definition, it is about what many call "food sovereignty" in the face of power and structural inequities. Alkon and Agyeman define food sovereignty as "a community's right to define their own food and agriculture systems" (8). Even that term is evolving. As my current community partner Fatuma Emmad of Frontline Farming told me recently, she no longer uses the term "food sovereignty" because of the connection to "sovereign," as in ruler, king. Rather, she now uses the term "food liberation."

Food justice connects to urban planning, environmental planning, health-care access, disability justice, affordable housing, transportation, and a whole host of other issues, which is why it is critical to view it as a systemic issue, not simply one of choice. Kimmerer invites us into not just a way of eating but a way of being:

> I cherish the notion of the gift economy, that we might back away from the grinding market economy that reduces everything to a commodity and leaves most of us bereft of what we really want: relationship and purpose and beauty and meaning, which can never be commoditized. I want to be part of a system in which wealth means having enough to share, and where the gratification of meeting your family needs is not poisoned by destroying that possibility for someone else. I want to live in a society where the currency of exchange is gratitude and the infinitely renewable resource of kindness, which multiplies every time it is shared rather than depreciating with use. ("Serviceberry")

Perhaps, through Kimmerer's lens, we can revise Pollan's suggestion that we move from what is "good to eat" to "good to think" to, instead, what is good to *be*. For her, that *being* cannot be done individually. It only makes sense in relationship, in gratitude, in kindness. Through all of these discussions of language, affordability, access, sovereignty, and liberation, the assumption among many local food justice advocates stands that desire for local food (whether it is accessible or not) is values based, even if those values may differ to suit each locale and culture. The abundance of natural gifts, languages, and values through which to understand and engage with local food signals an inextricable interconnectedness among beings and the earth that gives us life.

INNOVATIONS AND RECLAMATIONS IN LOCAL DEFINITION BUILDING

Based on these values, innovative programs have sprung up in every state in the country, and each one adds its own spin on what a definition of local food should include. Alkon and Agyeman claim that the food movement had been "itself something of a monoculture" in that it has consisted primarily of like-minded, well-to-do white people. They suggest instead a polyculture approach—lots of people and lots of different kinds of projects. This is what is happening more and more. From the Detroit Black Community Food Security Network farming abandoned lots in the city and Detroit's D-Town Farmers (*Urban Roots*) to Oregon's Seed Room CSA to Michigan's Ziibimijwang Farm selling maple sugar as an act of Indigenous food sovereignty, communities are creating for themselves a "critical food source for food-insecure populations" and reclaiming their power to control their food (R. White; Gottlieb and Joshi 148; Nelsen). Through shipping container vertical farms in Harlem and Brooklyn such as Harlem Grown and Square Roots, inner-city aquacultures like Denver's Growhaus, culturally relevant community gardens and farms like Black Sanctuary Gardens in Oakland and Frontline Farming in Denver, communities are bypassing industrial food and financial restrictions by growing their own food.

Some programs, such as Village Market New Columbia in Portland, Oregon, transform liquor stores with limited, industrial food options into healthy food and local food markets. Nuestras Raíces in Holyoke, Massachusetts, has taken a holistic approach to food sovereignty, creating a greenhouse, a restaurant, a commercial kitchen, a meeting space, a bilingual library on health and agriculture, many garden sites, and an education center. Eugene, Oregon, created a regional food hub through Hummingbird Wholesalers, and more regions are creating food hubs and processing centers because "small- and medium-size farms that choose to operate outside of the industrial system often lack the logistical tools necessary to gather, store, and transport food on a scale larger than a farmers market" (Grace 68). The Rosebud Sioux Native Nation manages a massive bison restoration project. The Sustainable Food Center in Austin, Texas, has weekly recipe swaps and cooking classes for WIC recipients. At the end of each weekly class, participants get a free bag of ingredients for the recipes.

Grassroots Gardens of Western New York is updating seven of their gardens to make them accessible for gardeners with physical disabilities by installing "wheelchair mats, seating connected to raised beds, and signage with QR codes (for people who use screen readers). They purchased accessible hand tools and garden scoots. They also created a long-term plan for enabling universal access in all their plots and committed to offering American Sign Language (ASL) interpretation at all public meetings" (Ferrante). All of these examples show community-led projects and programs that are rhetorical embodiments of how to define "local food." It is not about industrial excess at supermarkets or packed-in feedlots or uniformity. It is about cultural and community reclamation, connection, and accessibility. Each program mentioned here is an embodied example of *distributed definition building* in that each one "writes" its definition of what "local food" should mean through the program's embodied or written priorities, whether that be that racial justice must be included, that cultural sovereignty must be included, that skills development must be included, that access must be included, and so on, in the definition.

To combat industrial food's homogeneity and to encourage what Vandana Shiva calls "seed sovereignty," many farmers are bringing back heirloom and heritage seeds and meats, and towns across the country offer seed swaps, seed libraries, and seed banks (*Seed Sovereignty*). Ross Rodgers, founder of the Living Seed Library in Boulder, and one of Colorado's key seed savers, works to get seeds to everyday people both to inspire connection with the land and to build community resilience. His work to save and share seeds helps to expand what local food production is possible and, therefore, is an example of how each vital part of the locale's work on a social or environmental issue shifts a definition—in this case, the definition of "local food" expands to include access to seed and seed saving skills.

In a 2020 interview I conducted with Ross for this book in his backyard overlooking a creek and abundant gardens, he explains, "If we can save and breed seeds, actively selecting them to do well with less inputs, they will do better in more challenging conditions of drought or infertility. Growing out and saving seeds each year offers natural selection in environments that are a little bit more challenging. They will be resistant to challenging years, challenging climate changes, challenges of shorter frost-free growing seasons." I ask him what led him to create the Living Seed Library and other seed libraries in the region. He says, "We develop a relationship with the seeds. We can share the seeds and each seed has a story. A seed library is kind of another level to a seed bank, where the seeds are available to the general public." As libraries function as public places of knowledge exchange and literacy, Ross wants seed exchange and gardening to become normal and commonplace literacies again. He explains, "I want a resilient culture around me, because if people are able to feed themselves, if people have a basic knowledge about these things, then any kind of challenges we face in a food system, then people have something in them already that can come out when the need arises. I just know there's a lot of people out there who have had one little experience with growing because of what I've been able to offer. If they ever need to feed themselves in the future, it won't be a total mystery" (Rodgers). Seed banks and libraries like Ross's have popped up across the country, and according to Ross, this work is connected to a local food system that is defined through, for example, relationship, sharing, storytelling, preparedness, and resilience.

As Ross lists the names of a few of the hundreds of seed banks and libraries in the country—Rocky Mountain Seed Alliance, Native Seed Search, MASA Seed Foundation, Seed Savers Exchange, Seeds for the People, and Vibrant Earth Seeds—he clarifies how important they are in localized work toward seed sovereignty and cultural foodways: "A lot of Indigenous communities have lost their seeds or need to connect with the seed keepers in their region. Young people stopped farming. Land rights issues. They often don't have their seeds anymore. There's other Indigenous communities, other seed-saving organizations, that have saved a handful, or a significant amount, or grown them out into substantial amounts, and they're now bringing them back to the people that originally grew them" (Rodgers).

In addition to their work preserving heritage seeds, Black, brown, and Indigenous growers and producers all across the country lead the movements against corporate capitalist control of food and seeds and make explicit the systemic and racist legacies connected to food apartheid. For example, as Winona LaDuke explains,

> As colonizers drove Indigenous peoples from our territories, we were cut off from access to traditional foods. Starvation and disease became rampant. The forced reliance on inadequate government rations, often called "commodity foods," only changed the starvation from quick and obvious to hidden and slow. Today, Indigenous communities are recovering agricultural traditions linking past to present and future—and, in the process, restoring spiritual practices related to foods, while strengthening community health and self-determination. (*Recovering* 191)

Today's Indigenous-led fights against corporate control of land and water, as manifested by fracking and oil pipelines with the associated degradation of Native food sources such as salmon and wild rice, are parts of the ongoing struggle against settler colonialism.

A powerful example of work to break from a structurally racist food system and reclaim cultural heritage is at Soul Fire Farm in New York. As Soul Fire founder and farmer Leah Penniman explains in *Farming While Black*, part of the justice work of farming is to "heal our relationships to land" (8). The historical work she has done to connect the current food system to its roots in slavery and Jim Crow highlights the richness of the African and enslaved Africans' farming traditions. As Penniman so beautifully writes of her Black Latinx Farmers Immersion project at Soul Fire,

> We took turns cooking the recipes of our ancestors, substituting locally grown vegetables for their tropical equivalents. We learned the songs and prayers used in the process of slaughtering animals. We learned to take life. Then we engaged in herbal healing baths in the African tradition to cleanse that strong energy and lay down our metaphorical "knives." We used drums and songs to encourage the seeds to grow, and we filled the moonlit sky with the sounds of our dancing to Kendrick Lamar and Nicki Minaj. We bathed ourselves in resiliency. (*Farming* 8)

In addition to a reclamation of relationship with land among Black, brown, and Indigenous farmers, veganism is another form of cultural sovereignty that represents a return to health defined through traditional cultural foodways. The organization 10 Million Black Vegan Women, for example, calls itself "a groundbreaking public health intervention that uses plant-based nutrition and community support to address a preventable health crisis among African American women" (website). The website uses medical facts about systemic inequalities in the health system to energize Black women to change their and their families' diets and to understand misleading food labeling. In 2016, Pew Research "found that eight percent of Black Americans identified as strict vegan or vegetarian, compared with three percent of all Americans" (Rao). Black vegan bloggers, cookbook authors, influencers, and restaurateurs abound: from Vegan Voices of Color to *Sweet Potato Soul* blogger Jenné Claiborne, from *Sistah Vegan* blogger A. Breeze Harper to Cametria Hill's *A Southern Girl's Guide to Plant-Based Eating*. In Denver, hip-hop artist DJ Cavem's vegan-focused music and nonprofit emphasize environmental activism, local food justice, and urban gardening. When I asked my catering partner for several events that I put on in Denver, Johnitta Medina, the impetus for founding her plant-based soul food catering company Momma Jah's, she explained, "Exploring nutrition led me to a whole food plant-based approach not knowing veganism would provide a tool to further remove myself and my bloodline from the intentional, systemic harm against Black and brown bodies via the American food system. The land and her abundance has always been a source of healing for my ancestors. I am grateful to have been able to reconnect to that source-based power and knowledge through the power of plants" (email). These are only a few examples of a surging vegan movement among people of color who are defining healthy and culturally significant food choices for themselves. The rhetoric is focused on health, food sovereignty, and justice for Black and brown people, who are disproportionately made sick by industrial food.[2]

Eating connects us to natural cycles as well as to political and cultural movements; it can be an affective act, a rhetorical act, and a collective act.

---

2   While I cannot in the scope of one book name all of the projects of cultural food and health reclamation, I encourage readers to delve into areas that most interest you. I want to acknowledge that while I'm tracing certain threads of language, there are so many areas that I cannot fit into this book. There are fascinating studies on water, on plant-based movements, on diet culture, and much more.

The synergy of these concepts—the natural, the cultural, the embodied, and the collaborative—is where local food strikes a deep chord. The knowledge-based, culture-based, and values-based aspects of the food movements to move away from the industrial food system that are sweeping the country help illustrate how eating and food are discursively constructed.

Teaching *Ecological Community Writing*: Where to Begin

There is no single local food movement; each develops as it is needed in each locale. Because the definitions, goals, and outcomes are multiple, there is an ongoing rhetorical transformation that occurs as communities sow local food movements. And there is a discursive richness to each iteration of the movements. If there is no one way to instantiate or define a local food movement most effectively, how should I as a rhetorician and teacher understand my role and my students' roles in helping to create meaning or to influence the definitional discourse in our particular locale? It is this question that I pursue in the chapters ahead that focus on my local food work with students and partners.

Ecological community writing encourages an understanding of how language and writing are used across multiple scales from local to national to international and from multiple perspectives from personal to community, and beyond. This is because broad, systemic issues intersect in ways that will inevitably impact the local community writing that we, our students, and fellow community members do. In my writing courses on food rhetorics, before we turn to local manifestations of issues, students develop an understanding of how those issues tie to or respond to the industrial food system in the United States and elsewhere. When we study and teach how any issue manifests locally, we can begin by studying with our students how language about that issue is used nationally or globally, and how language, definitions, and rhetoric tie to systems of oppression or liberation. My students study, research, and write about the rhetorical choices authors, activists and organizers, farmers, government agencies, and companies make for or against certain foods, ideas, and lifestyles. The broader perspective introduces students to the idea of an ecology of writers, activists, companies, organizers, food workers, farmers, and nonprofit and government agencies involved in building definitions through which the public understands food.

Because course readings and students' research about industrial food can raise complex emotional reactions from anger to disgust to confusion, by about four weeks into the semester, after we've finished analyzing the national discourses and definitions, students are usually craving some ideas for how to move forward given all they've learned. How, they often wonder, can they individually have any impact against such enormous systems of inequality and power? That is why our turn toward how the national issues manifest locally, as well as local innovations, is so fruitful.

2

What's "Local" About Local Food?

REHINGING WORDS TO MEANING

Mirroring the national confusion over definitions of terms related to food, the rhetorical concept of "local food" shows up in sometimes generative and sometimes problematic ways in my home county of Boulder County, Colorado. In the pages that follow, I move from an exploration of national issues around definition building to local context. I offer my public scholarship and community-engaged pedagogy about the local food movement in Boulder County to show the discovery process at each iteration of the projects with students and partners and the new questions or barriers that became clear at each step. Community engagement work is so experimental and iterative that it would not feel possible or just to present a methodology and method without telling the story behind why and how they came to be. In writing this book, I knew it would be essential to explain how my own and my students' discoveries and ecological engagement, as in finding and building connections within the local food ecology, led to new community writing projects and questions. The book demonstrates the idea that we are building an ecology through our collaborative work, and this happens in nonlinear ways to continually address new questions.

https://doi.org/10.7330/9781646427208.c002

As I describe *ecological community writing* over the next several chapters, I explain how my method for multisemester and multiyear *distributed definition building* projects developed organically given the particulars of the local context. This chapter explains the beginning stages of my journey with my undergraduate writing students into understanding the importance of *distributed definition building* as we uncovered and then sought to understand the implications of widely differing definitions for "local food" here in Boulder County. Through analysis of several hundred interviews with Boulder County farmers and restaurateurs, stories in local media, and canvassed responses from farmers market patrons gathered by students in my writing classes through undergraduate action research, I reveal contradictions and confusion over the assumed meaning and value of the term "local food" leading to policy and business practices that do not always align with consumer desires and priorities.

In Boulder County, production capacity and demand-versus-supply issues as well as misunderstanding have led to complications in the definition of what constitutes "local food"; and, the various definitions used, often written by people in positions of power, are sometimes at odds with what consumers believe that they are purchasing. As City of Boulder and Boulder County government officials and residents consider how to build resilience into the food system, various groups vie to define the criteria for what constitutes "local food." Boulder County is not alone in these rhetorical challenges. Local food movements across the country now face both the challenge of and the opportunity for teaching community members to define what it is they want to support in food production and consumption. This articulation, as so many local food access and food sovereignty projects around the country demonstrate, does not have to be connected to power or economic wealth.

Moving from the national to the local interrogation of the intersections between food, language, and power, I urge rhetoricians to expose and begin to mend the rifts between words and their meaning, to help frame the public discourse and policy discussions around food. More broadly, I hope that my study of the City of Boulder and Boulder County's significant work to localize the food system will encourage scholars, teachers, students, staff, activists, and organizers interested in public rhetoric and community writing not only to educate the public on contentious social issues but also to encourage literate action that develops out of inquiry and emotional connection through *ecological community writing* using *distributed definition building*.

What's "Local" About Local Food—Miles, Ethics, Relations, or a Complexity of Factors?

Boulder County may be ahead of the curve in terms of the general population's awareness of some food-related issues; however, a type of "illiteracy" (Schell 81) persists around what "local food" means, and it is on this point of stasis that the rhetorical study I develop in this chapter focuses attention. The City of Boulder has long been touted as a liberal bastion and known by residents (as a vestige from its hippie days) as "the republic of Boulder" and "twenty-five square miles surrounded by reality." It buzzes with rhetorical energy around localization of the food system. A shared goal for the community, as expressed through government and nonprofit websites and media stories, has been "to meet [its] essential needs locally, and in the process to become more resilient and self-reliant" (Brownlee, "Local Food and Farming" 1). One of my interventions in Boulder's food localization movement has been to help "educate people . . . about the origins and contents of their food and about the systems that they support with each purchase" (House, "Re-Framing" 4). My study—corroborated through conversations, interviews, and work with numerous constituencies in the county to increase support for local food production and consumption—shows just how politically, economically, culturally, and emotionally complicated terms like "local food" can be, how tricky they are to define, and how vigorously people and the institutions they represent will fight to claim definitional ownership.

Powerful forces in Boulder County have coalesced around food localization as a benchmark of the county's resilience in the face of climate change and natural disasters such as flood and fire. In 2008, Boulder County commissioners set a goal of increased production of local food on the county's 25,000 acres of Open Space agricultural land to 10 percent by 2012 (Brownlee, "Increasing" 7). Production of local food accounts for only about 2–3 percent of the state's consumption, however, meaning that 97–98 percent of the food consumed in Colorado is imported from out of state (Brownlee, "Increasing" 8 and *Sustainovation* 5). County numbers may be even lower. According to Boulder Local Food Shift cofounder and independent scholar and author Michael Brownlee, of the $900 million of food consumed in the county, only 1–3 percent was produced in the county itself (Brownlee, "Thinking" 12). Clearly, production capacity would have needed to grow significantly to meet the commissioners' mandate. While still not close to the 2008 commissioners' goal, local food is one of the fastest-growing sectors of the Colorado Front Range economy, with community supported agriculture shares (CSAs), farmers markets, farm-to-table dinners, food-focused nonprofits, back- and front-yard gardens and farming projects, school garden programs, and food access organizations that have relationships with local farmers.

Boulder County has a strong though controversial record of supporting local agriculture, reaching back to colonization in the 1800s through "land claims" and erasure of Native cultures and foods, as I mention in the "On Lineage" section. As I quoted in the book's opening pages, the City of Boulder does well in its 2022 Staff Land Acknowledgment to offer recognition "that those now living on these ancestral lands have a responsibility to acknowledge and address the past. The city refutes past justifications for the colonization of Indigenous lands and acknowledges a legacy of oppression that has caused intergenerational trauma to Indigenous Peoples and families." This acknowledgment, if backed with ongoing action and repair, is one critical step because it counters troubling and persistent colonial rhetorics that I share in the pages that follow.

Boulder County government's rhetoric and practice have also shifted in important ways during the years I have been writing this book. According to the Boulder County government website, the County Open Space is used to grow primarily alfalfa and grass, wheat, barley, corn, sugar beets, pinto beans, and sunflowers. The types of production include conventional, certified organic, regenerative ag, organic practices, and nonorganic. First adopted in 1978 to "eliminate sprawl and to preserve agriculture, forestry, and Open Space land uses" (Boulder County Open Space), the updated 2020 Boulder County Comprehensive Plan reads, "[P]reservation of our environmental and natural resources should be a high priority in making land use decisions" (1).

While the county government website highlights sustainability in its promotion of local agriculture and is currently doing a tremendous amount of trailblazing work to support sustainable and regenerative agricultural practices, the county's historical definition of sustainability has not always been as easily understood. For example, the county commissioners voted in 2009, after highly contentious public hearings, to permit genetically modified (GM or GMO) crop production on leased Open Space lands. Some county government officials argued that they were considering the "sustainability" of the farm families, some of whom have farmed the land for several generations leading back to homesteader colonizers. After the 2009 decision to allow GMO corn and sugar beet production on Boulder County Open Space, one county commissioner at the time explained that it is "important to keep farmers on the land to act as stewards of that property. Therefore, the county needs to be careful not to adopt policies that would drive farmers away" (qtd. in Snider). "We need to keep farming this land," he added. "If we don't, it goes to weeds. It becomes unproductive, and we lose a piece of our heritage. We don't want to take an action that either abruptly or slowly chases farmers away, and we're left with fewer and fewer options" (qtd. in Snider).

The commissioner's rhetorical argument about maintaining the county's farm culture and the generational/heritage pride of stewardship perhaps conceals an economic reality—the county depends on revenue from leased land. It is also fraught to uncritically suggest a problem with losing a piece of white settler colonial heritage and chasing people away (as happened to the Native peoples in the most violent of ways) from land. Native heritage and foodways that could offer alternate ways to live with the land and produce food were not discussed. An important part of the county's reckoning with the past's ongoing legacy is to address and repair the rhetorics of "agricultural heritage" that have been dominant. For example, the county website still links to a 2006 document, "Boulder County's Agricultural Heritage," which begins its historical study in 1859 with "Early Settlement/Pioneer Agriculture" (Wolfenbarger). The only passing mention of Native foodways in the document illustrates the author's apparent lack of interest in foodways before white settler arrival:

> In just a few decades, Boulder County went from an Indian hunting ground covered with prairie grasses to bustling mountain mining camps supported by successful farms on the plains. Settlers arrived shortly after the discovery of gold, broke sod, established farms and ranches, organized and built irrigation systems, founded farming communities, and organized communal agricultural societies and county fairs—all in less than thirty years. (3)

The author's emphasis on rapid "progress" ("all in less than thirty years") without any recognition of the violence of settler colonization was common in discussions of the early years of colonization in the area.

In its discussion of local food and sustainable agriculture successes, when I began writing this chapter in 2017, the county listed important programs such as Double Up Food Bucks at farmers markets and grocery stores for increased access, and its multiple programs to monitor and improve soil health, increase cover crops, increase drip irrigation, develop carbon sequestration projects, and *preserve agricultural heritage*. It is this final element of "agricultural heritage" that landed us in problematic waters around issues of sustainability. First, *whose* agricultural heritage are we working to preserve? The county was not talking about Native food cultures. The concept of sustainable agriculture practices holds different meanings for different constituencies and cultures, and as far as I am aware, only white settler family heritage and agriculture were considered in these earlier discussions. Words and rhetorical choices have profound implications and can perpetuate violent settler colonial legacies of erasure. Currently, in 2024, the language of "agricultural heritage" has been removed from the county website, and a county government webpage shows County Sustainable Food and Agriculture Fund grants awarded to many recipients, including the Native-led organization Harvest of All First Nations.

I'd like to zoom in on the GMO issue in particular, as it can highlight the importance of how sustainable food and agriculture are defined. As I've been writing this book, county policy about growing genetically modified crops has shifted back and forth. Public pressure initially led to the county's reversal of the 2009 decision with a mandate that farmers phase out all GM crop production on Open Space lands. In December 2021, however, Boulder County commissioners announced that they would no longer require the phaseout of GM corn and GM sugar beet seeds. As in 2009, this decision was preceded by contentious public debate over what sustainable local agriculture means, with pro- and anti-GMO advocates arguing that their side was most sustainable for a variety of reasons. As of the writing of this book, the Boulder County website page titled "Local Food and Sustainable Agriculture" states:

Support for Agriculture
The county will encourage the preservation and sustainable use of agricultural lands as a current and renewable source of food and feed and for their contribution to cultural, environmental and economic diversity.

Local Food Production
The county will encourage and support local food production to improve the availability and accessibility of local foods and to provide other educational, economic and social benefits.

Sustainable Agriculture Practices
The county will promote sustainable agricultural practices on publicly owned lands and will encourage them on private lands. Sustainable practices include production methods that are healthy, have low environmental impact, [are] respectful to workers, are humane to animals, provide profitable agriculture opportunities to farmers and support farming communities.

As this information from the county website indicates, supporting "sustainability" may mean different things to different people: environmental sustainability, organic and regenerative farming practices, economic sustainability for commodity crop farmers, or sustainability of farming, including GM crop farming, for families who have farmed the land for generations. The county states in its 2021 "Boulder County Parks and Open Space Cropland Policy," "Sustainability is one of Boulder County's Guiding Values," which it defines as "the use, development, and protection of resources in a way that enables Boulder County residents to meet their needs and maintain a high quality of life, without compromising the ability of future residents to do the same" (4). The economic needs of conventional farmers who grow GM crops are clearly important and reasonable to consider; however, many county residents and organic or regenerative farmers argue, allowing for GM crops is not necessarily synonymous with comprehensive, long-term sustainability, which requires consideration of the unintended consequences of the farming practices themselves.

In fact, many of the most progressive farmers in the county argue that we should not aim for "sustainability" of how things are but rather for regenerative agriculture that helps to grow back (re-generate) our topsoil. Indeed, "regenerative ag," currently listed among the kinds of agriculture practiced on county-owned Open Space, is a more recent addition to the list and accounts for the county's commitment to projects that regenerate soil health, in particular through Boulder County's Restore Colorado program, a multiorganization initiative that creates "an economic connection from consumers and businesses back to farmers and ranchers, soil health can be restored and carbon can be sequestered to improve the overall resilience, prosperity, and nutrient density of local land" (Boulder County, "Restore").

Across the United States, conventional farmers, whether or not they grow GM crops, typically use synthetic pesticides and herbicides, sometimes including the neurotoxin glyphosate, the main ingredient in Monsanto's product Roundup. In 2015, the International Agency for Research on Cancer listed glyphosate as a "probable human carcinogen" (Landrigan and Benbrook). GM crops are specifically designed to be resistant to particular pesticides and herbicides, including glyphosate, and therefore GMO farming often includes the use of this toxic chemical. In addition, the tendency of conventional farming practices to lead to progressive degradation of soil and water quality is another factor that has led many critics to call current GMO farming practices unsustainable. In a critical decision in 2023, while continuing to allow production of GM crops, Boulder County banned the use of glyphosate on county-owned lands ("Invasive").

The issues and political considerations here are not simple and straightforward. With many stakeholders at the table (and many excluded), and with people's livelihoods at stake, sustainable or regenerative real-world solutions can be challenging to achieve. However, in determining to allow the continued farming of GM corn and sugar beets, Boulder County has made a decision that, based on the heated public debate noted previously, many of its residents find inconsistent with their own definition of "sustainable" as well as with the county's stated commitments. The definitions, then, for what "sustainable local food and agriculture" production means and how to support it vary widely.

Michael Pollan's argument about the misuse of words and the need to realign them with "real things and precise concepts," which I discussed in the previous chapter, resonates here regarding the county's definition of the term "sustainability" and necessarily connects to their use of the word "local." Because industrial agriculture's reliance on synthetic pesticides and herbicides is not indefinitely viable in terms of environmental impact and public health, the use of the word "sustainable" in this context unhinges the word from a clear meaning. Are commodity crops grown or produced in Boulder County but slated for export out of the state for use in processed, industrial food included in the county's understanding of "sustainable agriculture"? The answer will depend on whom you ask. We then come to the definitional question guiding this chapter: How is "local food" to be defined, and who gets to make that decision?

Community-Engaged Undergraduate Writing Research

A course framed through an *ecological community writing* methodology can begin with students analyzing texts and conducting research about national rhetorics and definitional conflicts around an issue, as I discussed in the previous chapter. Through this work, which in my courses takes about four to six weeks of the semester, students research complex systemic issues in food literacies while developing critical thinking, writing, and rhetorical skills. We then turn to ways in which the issues manifest locally. Prior to 2014, I had been focusing community-engaged project development on one discrete local writing project per class each semester developed between me, my students, and my partners (House, "Re-Framing"). While this model of community-engaged writing often produced intended results for our partner and enhanced my students' learning and engagement, it followed a rather traditional "critical service-learning" trilogy moving students through the questions "What? So what? Now what?" (Mitchell; House, "Reflective").

All of this changed in 2014, when a community project propelled me into research of an ecology of local food extending through multiple sectors including farms, restaurants, consumers, schools, grocers, businesses, nonprofits, and farmers markets. I worked on this evolving community project until I left the University of Colorado Boulder for a faculty position in the University Writing Program at the University of Denver in 2021. While I still live in Boulder, I am slowly implementing *ecological community writing* in my new position, working anew to familiarize myself with place and an ecology of new partners and work in Denver.

The shift in my pedagogy began in the spring of 2014, when a nonprofit organization in Boulder asked me to design a community-engaged writing course in which students would conduct interviews with farmers and restaurant chefs or managers, the questions for which were written by the nonprofit. The goal was to allow the organization to assess how food distribution from farms to restaurants was disrupted by the massive September 2013 floods in Boulder County. This was the research question with which my students began. Because I will be discussing several years' worth of student projects in the pages that follow, to avoid confusion I will call this the interview semester.

Interview participants were able to choose between a phone, email, or in-person interview. After the project's approval, students conducted and recorded the interviews and wrote them up for the nonprofit. Though the interviews were not originally conducted to reveal problems concerning local food definitions, as I read through the student interviews of 110 farmers and restaurateurs as rhetorical artifacts after the project ended, I noticed an interesting rhetorical problem.[1] I discuss it in the following text not to shame or blame any particular restaurant—and I will not name any restaurant, chef, or manager—but rather to lay out the complexities and confusion as a starting point for rhetorical intervention and to highlight the ways that I encourage Rhetoric and Writing practitioners to intervene in public discussions around contested terms.

Many Boulder County restaurants that claim on their menus to source local and organic ingredients "whenever possible" actually purchase little local food. As Lisa Mastrangelo says of cookbooks, so too with menus: they can be read "as rhetorical artifacts that reveal much about their communities" (73). I was particularly interested in the interviews from thirty-four nonchain, locally owned restaurants. A few of these restaurants source no local food. Three managers gave responses that exemplify answers common in this category of restaurants: "[S]ustainability is not of big importance to us"; "I have no knowledge of where the food comes from"; and "[W]e can't afford to sell locally-grown food." One manager cited Sysco, the largest food distribution corporation in the country, as the farm from which they source.

The majority of the nonchain restaurant interviews revealed that restaurateurs are at least aware of the local food trend but are struggling with how to adapt due in large part to cost and supply. Three main thematic concerns emerged in the interviews:

---

1   I want to emphasize that this was not a formal research study. Once I saw the interesting rhetorical problems in some of the responses, I contacted the Institutional Review Board (IRB) office to ask how I might use the interviews in further work and publication. I was assured that as long as there was no identification of restaurant, restaurateur, farm, or farmer, the way I am using the information is not generalizable and can be used without IRB approval.

1. Restaurateurs do not know how to or cannot source local food when the Colorado growing season ends.
2. Restaurants use misleading language, which confuses consumers.
3. Restaurateurs do not agree on what constitutes "local": Is it miles, ethics, relationships, or a combination of several factors?

When the Colorado harvest ends, or when distribution is impeded, most restaurants offering local produce during the season are more likely to turn to large distributors such as Shamrock, Sysco, and US Foods for most of their ingredients. One group of restaurants shares a farm to grow produce, but according to the owner (who has stated this figure publicly as well), they can only grow about 10 percent of the produce they need for the restaurants. One restaurant has its own herb garden, but all other products come from Sysco. Two restaurants purchase exclusively from a large-scale distributor, except for an occasional one or two local items on special. In other words, the "whenever possible" of menus' claims is not always what consumers might imagine.

One chef told students, "During winter months it is very hard to source locally, and we mainly go through large purveyors." Another answered that it is "difficult to find a reliable source of local and organic meat, though customers have been pushing for it." Conversely, another restaurateur worried that customers would not want to pay extra for local and organic food: "We do our best to stay organic but want to provide customers with affordable food." One manager of a restaurant that makes claims on its menu to be farm-to-table told students that they would not answer questions about sourcing (particularly meat), as the restaurant's ethos would be damaged if consumers were aware of their purchasing habits. Several with the "we source local" language on their menus claimed to source from a certain number of farms, one restaurant even sourcing from as high as fifteen farms, but the manager could not name a single one. Several restaurants have implemented important "sustainability policies" that include composting, recycling, and non-Styrofoam to-go containers, but sourcing sustainable food is not part of the restaurants' sustainability paradigm. How each restaurateur and consumer defines "local food" may be vastly different.

These findings signal a possible disjuncture between consumers' considerations for purchasing locally sourced food and the restaurants' availability, cost, and convenience considerations.

As Gottlieb and Joshi report, surveys conducted by both the Food Marketing Institute and by the Restaurant Industry Association (now the National Restaurant Association) indicated that about 75 percent of shoppers and restaurant patrons regularly purchase local food and are more likely to frequent places that offer local food (183–184). But marketing often confuses patrons as much as it informs them. In fact, "Miele and Evans found that labels carrying [animal] welfare claims created two distinct outcomes... The first effect was that it created an 'ethically competent consumer' while the second effect was opposite—creating an 'ethically noncompetent consumer'" (cited in Abisaid 146).

This happens with local food claims as well, as evidenced in several studies and exposés. For example, in 2016, investigative journalist Laura Reiley uncovered several seafood restaurants in Florida that boasted "local" seafood actually deriving from all over the world. In a study of greenwashing by Tony's Market, Megan Koch and Cristin Compton note that employees did not have a shared understanding of the term "local"—all of the respondents' answers centered on distance, not sustainable practices or treatment of workers. The marketing did not explain the benefits of local products either. Koch and Compton conclude that "greenwashing can discursively close off communication between a corporate entity, its employees, and its customers" (243). These stories highlight two distinct problems in "local food" advertising—deliberate false advertising and lack of understanding. While not all of the confusion is insidiously or intentionally sown, it does indicate that businesses have been able to appropriate the language of "local" for their purposes.

Four chefs that my students interviewed could be considered activist chefs in that they attempt to enact their ethical considerations through their restaurants' practices. One chef describes his enterprise as a "full-circle restaurant, committed to using every part of the animal and plant." He elaborates: "[L]ast summer, much of the compost from the restaurant was taken back to the farms from which [they] source [their] meat and went directly to feeding the animals." Another explained, "Sustainability means understanding that food sourcing should be as close to staying within the foodshed as possible. Local means not industrial, small, within the county." One said that local food has to be "non-GMO." In one restaurateur's discussion of "local," they explained how important relationships and community are: their restaurant bought hundreds of pounds of squash from one farm during the 2013 floods so that the farm's produce wouldn't go bad, and their chef changed the menu accordingly.

One restaurateur at a farm-to-table restaurant urged the student interviewers to think beyond a miles-driven definition of "local." At this restaurant, they deliberately forge relationships with small-scale producers around the globe—berry farmers in California or salmon fisheries in Tasmania—to support what they define as ethical practices. Interestingly, the restaurant owner, who "refuse[s] Sysco products," defined these purchases as "local" because they feel that the small-scale, noncorporate businesses they support align with the ethics that consumers associate with local food. Similarly, in John R. Thompson's analysis of the Eat Local Challenge website, he notes their inclusion of fair-trade products: "Such goods are globally sourced and certified by certain organizations as products being grown, harvested, and produced with environmentally sound activities and fair labor practices. In that sense, global goods can qualify as at least a form of 'local' consumption" (66). The subsequent discussions my students had in class about this concept surfaced multiple perspectives and displayed differences in what students highlighted as important to them—valuing food miles, a variety of ethical considerations, relationships, or some combination of factors when defining the "local" in "local food."

Overwhelmingly, the farmers that my students interviewed for the same project cited "tight-knit relationships" as the most critical element in their work with restaurants. Several farmers travel to restaurants to bring samples to chefs or to drop off products, though doing so is time consuming. When one student asked about the impracticality of this time-intensive practice, the response reflected the notion that it is about relationship building. "[We're] so much more than just a list of crops," one organic farmer said. One regenerative (not certified organic) farmer mentioned his pride in producing "the healthiest, most nutrient-dense produce possible. No pesticides (not even organic ones), no herbicides, no government certification. We've been doing research into the connection between phytonutrients and human health, and it's exciting to see the role farms can take from variety selection to harvest to preparation techniques to maximize the availability of these cancer-fighting, health-promoting nutrients." Seven farmers used the word "pride" associated with feeding fellow community members food that is healthy for them and the environment, as well as eagerness to forge relationships with restaurateurs who share similar values. Only one farmer mentioned distance ("fifty-mile radius") in their definition of "local food." The restaurateur and farmer interviews exposed critical definitional inconsistencies around the term "local food" to indicate that people cannot presume either transparency in advertising or a shared understanding of the term's meaning.

The Method for *Distributed Definition Building* Begins to Take Shape

LOCAL MEDIA BIBLIOGRAPHY AND DISCOURSE ANALYSIS

In the semester following the student-conducted interviews, I taught a community-based upper-division writing course entitled Food and Culture. I wanted these students to delve more deeply into the definitional complexities materialized in Boulder County and City policy, practice, and rhetoric, so after reading about the national issues around language and industrial food, we studied the previous semester's student-conducted interviews as a launching point into our investigation of the local food terrain. A new research question emerged: Is there widespread confusion in the county about what constitutes "local food?" I asked the students how we might investigate the question.

They chose to create a shared annotated bibliography on Google Docs of ninety local media sources from 2010 to 2015 that focus on local food. These included newspaper articles (*The Denver Post, Boulder Daily Camera, Boulder Weekly*), radio shows (KUNC, NPR, CPR), blogs, nonprofit publications, local magazines, and social media sources. Students added their sources on the Google Doc rather than submit them individually so that we could think ecologically about the writing happening in the community—where is it happening; where is it not; who writes and has a voice in community discussions; who does not. The students then conducted a community discourse analysis to understand the multiple ways in which local food is discussed in their local media sources, including the important trends and any confusion or complexity of definitions that emerged. Students saw that as in the interviews conducted by my classes in the previous semester, the media sources revealed a similar definitional confusion over the term "local food"—is it about miles, ethical considerations, health, relationships, or something else?

Student interviews, annotated bibliographies, and community discourse analyses that focus on contested terms can be assigned for any number of social issues, and my methodology for *ecological community writing* begins with this preliminary research at the national and local levels so that students grasp the complexity of issues and the terms associated with them as well as the local ecology of who seems to have the power to create definitions. The assignment can give students a grounding in local conversations and definitions around key issues and terms depending on the thematic issues they are studying in a course. Neither my students nor I have definitive answers for how terms should be understood, which positions them and me as co-learners delving into questions of language and rhetorical choices together. As students each semester study a different part of the ecology related to the issue, they will help point to the multiple definitions that they find, look for gaps or discrepancies, and study their findings in relation to findings of students from previous semesters. Their work is helping to not only study but build connections in the ecology.

I conducted my own community discourse analysis of the ninety local media sources my students found. I did so to understand the dominant discursive ways in which local media discuss and define local food. My analysis corroborated my students' findings about the media sources, revealing differing perspectives on the concept of "local food." Interestingly, no source listed distance/proximity as the key definitional factor. Using the grounded theory method of open coding, I determined codes ("words or phrases that capture a central or notable attribute" [Arthur]) that related to the specific ethical, environmental, or cultural considerations that framed the texts. I then used the coded values to determine more specific thematic clusters that I saw emerging from the data (see table 2.1).

The local media discourse focused principally on benefits to the local economy and the environment, as well as on relationships and community building. The sources' discussions of local food were more positive than negative, with 39 percent emphasizing innovations in local food production, distribution, and education. Thirteen of the positive sources discussed K–12 projects such as gardens in elementary schools and curricular design around garden production. Other sources discussed innovative ways in which metro-Denver and Boulder are increasing capacity through vertical farming, controlled environment spaces, food forests, which are diverse

TABLE 2.1. Thematic clusters in ninety local media sources depicting the areas of focus.

Media source themes[a]	Sources that include the listed theme (%)	Sources that include the listed theme (n)
Sustainable food economies/economic benefits	37.8	34
Environmentally friendly/sustainable (mention relation to climate change, food miles/fossil fuels, soil health, or pesticides)	23.3	21
Interpersonal connections/community/support small-scale farmers	22.2	20
Tastes better/better quality/freshness	17.8	16
Self-sufficiency of the state/CO's food security	8.9	8
Healthier/nutrient-packed	8.9	8
Want to know how grown, where grown, and by whom	5.6	5
Ethics of meat production	4.4	4
Cultural food sovereignty	3.3	3
Political act/concern about democracy	2.2	2
Spiritual connection to food	1.1	1

a. Many sources listed multiple themes, and I counted them for every theme they list.

plantings of edible roots, plants, and trees for human consumption, and "cottage food laws," which allow Colorado residents to sell nonhazardous food products that they grow or produce at home; and increasing year-round access through preservation techniques, indoor markets, and aquaponics. Twenty-seven percent vocalized concerns related to supply, access, and the short Colorado growing season. The restaurateur and farmer interviews and the local media coverage together indicated a pattern of people making (sometimes) inaccurate assumptions, whether intentionally or not, about the varied ethics or benefits behind the concept of local food, however those ethics or benefits may be defined. The interviews and media indicated similar concerns about barriers, which my students and I noted as issues that might merit further consideration and intervention.

In 2009, the Food Marketing Institute conducted a national study asking people in the United States why they buy local food. The top three reasons listed were freshness (82%), support of the local economy (75%), and knowledge of where the product came from (58%) (Brain 1). I suspected that Boulder County consumers have reasons for selecting local products, in line with the themes that emerged in the local media discourse analyses and restaurant and farmer interviews, and that the consumers may not be aware of the variety of definitions the county and city governments, producers, and restaurateurs use when they state their commitment to local food. I wanted to determine whether consumers understood locale to be connected with particular ethical, moral, environmental, cultural, or other values. If so, what are those values? These were the research questions I posed to my next semester of students as the launching point for their final project, which they designed.

FARMERS MARKET LITERACY PROJECT

After reading the interviews with restaurateurs and farmers (interview semester) and reviewing the local media source documents (local media semester), my third semester of students wanted to investigate whether there was dissonance between the material and imagined realities of Boulder County's local food rhetorics. Recognizing farmers markets, as Justin Eckstein and Donovan Conley do, as "rhetorically charged civic space[s]" that "display the rhetorical machinery of ethics that animates contemporary consumers' desire to visit the market[s]" (172, 174), my students and I rented a booth at the Boulder Farmers Market, a producer-only market that accepts direct payment, Market Bucks, SNAP (Supplemental Nutrition Assistance Program), and WIC (Women, Infants, and Children), including a popular Double Up Food Bucks program that matches dollar for dollar SNAP purchases of fresh fruits and vegetables up to $20 per visit. For the class's community food literacy project at the market, students named their project LOCAL, Love Our Community's Agricultural Life, with a twofold goal: to share and to gather information.

We wanted to educate people about the definitional complexities of "local food" so that consumers would know to ask questions and not assume a fixed or shared meaning. I use "we" here deliberately because at this point in the semester, my students and I were genuinely discussing together ideas and project options that I had not preplanned. I let students know

FIGURE 2.1. Three students get our tables ready at the farmers market. On a table, they set up informational materials they had created for the event to build community food literacy. Also pictured on the tables are donations of produce and products from farmers and businesses at the market to be used in a giveaway. (Photo credit: Veronica House)

FIGURE 2.2. In the background, students with clipboards engage with farmers market patrons about definitions of local food. In the foreground is our market table with student-generated materials and donated goods. (Photo credit: Veronica House)

that they were co-creating the rest of the course with me. For this project, we spent a class period working on a catchphrase that embodied the learning my students wanted to offer market patrons, "Local: It's about more than just miles." This, they hoped, would make people think, "If it is not just about distance, what *is* local about?" That question, which distilled the issue into a succinct, hopefully memorable phrase, would then open up dialogue without forcing a prescribed definition.

The students canvassed 156 patrons over a five-hour period at the market to ask what their considerations are when they purchase local food (see figures 2.1 and 2.2). The prompt was "Define 'Local Food' and what it means to you." They introduced themselves and their project to market patrons, and after the marketgoers wrote their responses on index cards, the students

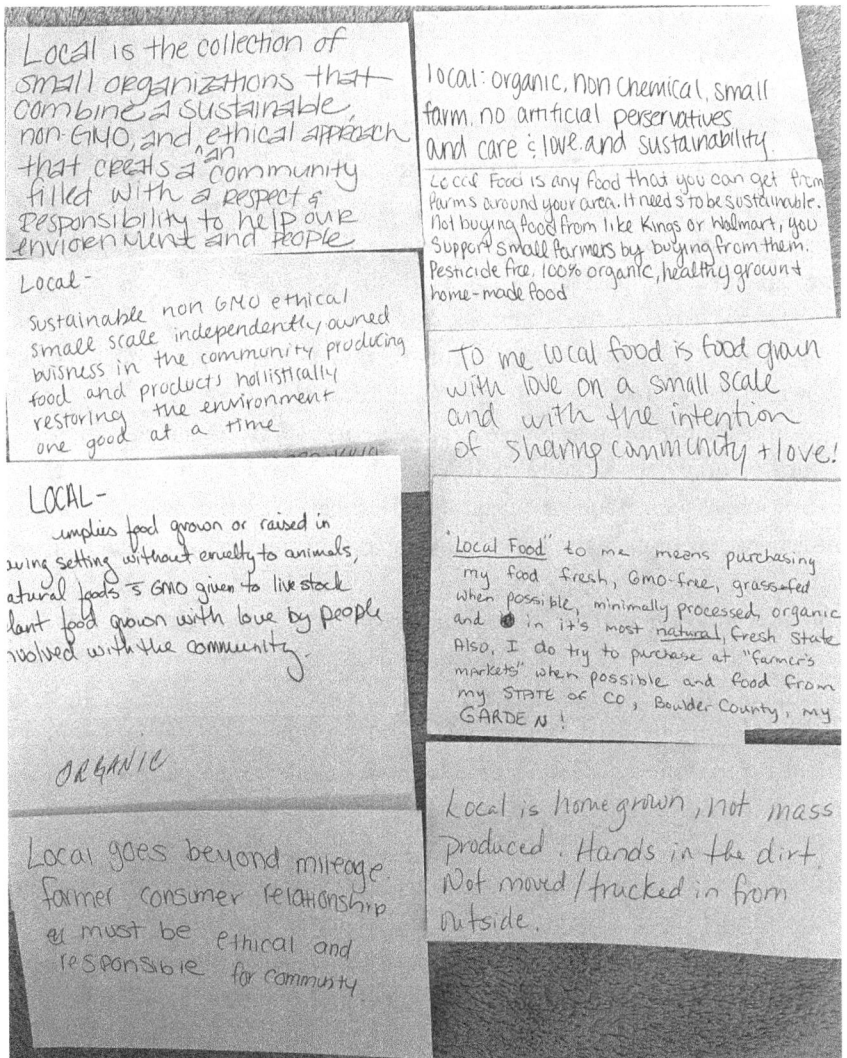

FIGURE 2.3. Samples of ten farmers market patron responses, handwritten on index cards, are an example of *distributed definition building* across a local ecology. (Photo credit: Veronica House)

engaged them in discussion, presented an informational flier and a refrigerator magnet that they had created for the event, and entered interested participants in a raffle of market goods that generously had been donated by farmers and businesses at the market in support of the students' investigation (see figure 2.3). Creating relationships in soliciting donations required genre and audience awareness and helped to build connections in an ecology. Students were both gathering information and engaging in a public food literacy project.

Certainly, farmers market patrons are a self-selecting group, and their very patronage seems to make it more likely that they consider food sourcing at least to some extent. While their responses may or may not be representative of the larger Boulder County population, they gave us a good indication of what people who deliberately seek out local food believe that they support through their purchases. As with the media sources, I used open coding in my analysis of the patron responses. And as with the media sources, patrons' responses revealed several compelling thematic clusters that indicate what they care about as well as their misconceptions (see table 2.2).

While the highest percentage, 46.2 percent, of patrons includes distance as one of many considerations but not the most important one for their definition of local food, environmental concerns as the most important consideration topped the list at 36.5 percent of responses. Given Boulder community members' proclivity toward progressive environmental action, this percentage is not surprising. Importantly, though, eight people wrote that "local" means non-GMO, and fourteen people wrote that "local" means organic. Given my research, this is clearly not always the case. (Indeed, at the Boulder County Farmers Markets, not all produce or meat is certified organic or non-GMO.) In other words, customers may sometimes make incorrect assumptions about the health or environmental quality of the food they purchase. This finding reinforced my students' belief about the need for a consumer-directed educational campaign to emphasize the importance of asking pertinent questions of producers and retailers.

Another source of confusion involved distance. Carbon footprint or food miles were frequently cited, but market patrons had no agreement on how close "local" should be. Responses ranged from "a bike ride away" to "within my town" to "50 miles" to "within a six-hour drive." This broad range in what patrons considered acceptable distances highlights how arbitrary any fixed distance is in determining a definition of "local food." Perhaps this is why fewer than half of the people my students engaged at the market listed distance at all as a factor for what "local food" means and why only 18.6 percent mentioned distance as the sole factor.

Responses overwhelmingly used positive language associated with environmental benefits rather than negative environmental impact, as in "the animals and crops are raised with care, using sustainable labor and

TABLE 2.2. Thematic clusters in farmers market patron responses.

Farmers market response themes[a]	responses that include the listed theme (%)	responses out of 156 that include the listed theme (n)
Distance as one of many considerations, but not as most important consideration	46.2	72
Environmental concerns	36.5	57
Supports community (jobs and local economy)	28.2	44
Relationships/know your farmer	24.4	38
Used word love, heart, and/or care	14.1	22
Mention distance only	18.6	29
Ethical concerns (workers, animals)	14.1	22
Supports small-scale business	10.3	16
Must be organic to be local	9.0	14
Must be non-GMO to be local	5.1	8
Taste/freshness	4.5	7
More healthy	4.5	7
Concern about future/big picture	1.3	2

a. Most responses listed multiple themes and are counted by every theme they list.

mindful of the local ecosystem" rather than "not big, commercialized companies." Two people framed their choice to purchase local food as tied to concerns about systemic and sustained issues: "Local is more than sustainable. It addresses the social issues at the root of our environmental crisis through community values and food security" and "Our actions affect the next seven generations. The Earth gives to us, so we should give back to the Earth with good compost, clean water, and restored soil integrity." From the responses, it is apparent that many consumers, at least in this setting, are aware of the connections between environmental issues, local economies, and food production.

The fact that people framed their responses positively in terms of environmental and relationship-focused benefits offered me and my students a clue regarding how to effectively influence people's views of local food and, perhaps, their purchasing habits. In other words, despite knowledge about negative environmental or economic concerns tied to industrial food, market patrons wanted to focus on the positive implications of their actions rather than fear, anxiety, anger, or other negative emotions. This revelation became significant in terms of how future student projects would use positive rhetorical frames for discussions of contested and complex issues, as I discuss in detail in the next chapter.

Particularly interesting in relation to these findings is the frequency of consumers' consideration of how local food supports community and relationship building, and who is included or ignored in that community and relationship. Over 24 percent wrote responses such as "I want to know the farmer on a first name basis," or "made close by, by people I know and trust," or "I can shake the producer's hand," or "brings people together with a sense of something worth doing to create a better world and protection of our precious earth." Fourteen percent of these responses specifically use the words "love," "care," or "heart," as in "grown with love by people involved with the community," "I buy local food because I love my community," or "local means friendly, together, community, face-to-face, heart-to-heart."

Gaps in Definitions of Local Food

No farmers market responses mentioned race, access, cost, or cultural sovereignty as considerations in purchasing or supporting the production of local food (though the one response mentioned previously references without attribution the ancient Haudenosaunee Confederacy's "Great Law," that "the decisions we make today should result in a sustainable world seven generations into the future") (Indigenous Corporate Training). Boulder is the seventeenth-richest city in the United States as of 2018, with a median household income of $85,364 (Stebbins and Sauter). It is 67.4 percent more expensive to live in, with grocery prices almost 10 percent higher than the national average (BestPlaces). The City of Boulder faces a poverty rate of 21.3 percent, almost double the national average of 13.1 percent, and an estimated 13 percent of people in the county experience some level of food insecurity (Boulder Food Rescue). My students were surprised that no reference to these issues surfaced, and they noted that Boulder has a façade of wealth that tends to hide communities in need.

These gaps at the Boulder Farmers Market mirror omissions in some of the countywide conversations about local food. Perhaps because, according to the latest census, Boulder County's population identifies as 76.2 percent white, 14.7 percent Hispanic or Latinx; 5.2 percent Asian; 1.3 percent Black or African American; 1.7 percent identifying as two or more races; and only 0.9 percent American Indian or Alaska Native, the most visible local food campaigns in Boulder County during the decade that I was involved did not tie access and justice to racial justice. Conversations at local food gatherings, workshops, and farmer events have been most frequently framed through rhetorics of resilience and farmer support—the ability to produce enough food for the population in case of crisis and the ability to produce enough local food to meet demand. Because of farm failure rates above the national average (as I discuss later in chapter 4) and the high cost and intensity of labor associated with small-scale vegetable production, farm viability and farmer support are dominant and critical issues in the county's local food discourse. A key element for many of my community partners in the local movement is to educate the public about the importance of supporting farmers (at the time of this writing, only one small-scale organic farmer in the county does not have two or more off-farm jobs) and understanding the high costs of production to grow the foods.

According to an interview I conducted for this book in 2020 with then–Boulder County Farmers Markets executive director Brian Coppom, the farmers markets play an important role in feeding our community, particularly through the Double Up Bucks program for WIC and SNAP customers. Brian estimates, "[L]ast year [2019] we did about $325,000 in food incentives that we distributed and redeemed at the market. This year [2020], our budget is closer to $500,000 to do that." While these numbers are from five years after my class's farmers market event, the program existed in 2015, and neither the program nor its benefits for our community showed up in market patron responses. The gaps indicated to me and my students that topics of access, racial justice, and affordability would be necessary to highlight in a community food literacy campaign.

Indeed, local food movement rhetorics in Boulder County have historically tended to be different and often separate from food justice and food assistance rhetorics in local organizations such as Boulder Food Rescue, Emergency Family Assistance Association, and Bridge House. While there are increasingly connections between these organizations and local farms that may donate produce, most of the organizations' focus is on food waste, affordable access, and culturally relevant food, and not as much on local food production and consumption. To be clear, all of the work mentioned here is incredibly important. I will turn later in the book to some of the more recent work in both the county and the City of Boulder to support a more culturally diverse local food movement and, specifically, to work with the members of several Tribal Nations on land use issues and educational projects. But there has historically been a rhetorical disconnect between the local food movement in Boulder County, which is predominantly a white-led movement that engages predominantly white community members in support of predominantly white farmers, and the food access and justice work happening in the county in support of a racially diverse population. Work to make these connections does happen, for example, in the Native-led cultural sovereignty organization called Harvest of All First Nations, but it has not been part of the dominant discourse. This dissonance between local food conversations and racial and cultural justice and access conversations is decidedly different from conversations in many parts of the country, including in Denver, where food access and local production discussions are often one and the same, bound together in purpose and often led by people of color in the communities where the food is being produced by community members in racially diverse neighborhoods throughout the city.

In terms of *distributed definition building*, in Boulder County, local food discussions are different from those in Iowa or Detroit or Denver. It is important to bring justice and historical erasures into discussions about local food production in Boulder County. I also hope that as residents and agencies of Boulder County work to make farming a viable business path for our currently predominantly white farmers, we will also commit to build opportunities for success and incentives for BIPOC people who would like to farm in Boulder County. As organic and regenerative farming become more supported through policy, I hope that these practices will become more common and affordable, which ideally means costs go down and more and more people can afford healthy, locally sourced food. As City of Boulder and Boulder County residents reckon with the history of agriculture here and what the previous discussions of the importance of recognizing heritage have implied, and as the city and county increasingly strengthen relationships with Native Nations members, I hope that acknowledgment of Indigenous foodways becomes increasingly woven into discussions of local food and land use. Soil regrows; our bodies become healthier; people renew their relationship to the land; we confront our locale's history in substantive ways. Change happens in systems and is not linear, and every small change matters to impact the whole.

Through my students' investigation into the complexities involved in Boulder's efforts to define and promote local food, they and I interrogated various rhetorical choices made by different local constituencies. Over an organic meal, sourced from our day at the farmers market, around my dining room table during one of our final class sessions, my students and I read through the farmers market patrons' responses together. We talked about the thematic patterns we saw as people defined local food for themselves. We noted what may have been missing from one response, what another response expressed well. After the farmers market project, the students wanted to publicize their findings (and the interview and local media findings of the previous two semesters of students' work) so that all of their work might be of value beyond our classes. They asked, even though the semester was ending, that I share the three semesters of work with people who could use their findings to make a difference. They entered a long-running community conversation and found information that could impact it. They knew that their knowledge-building work was not bound by the timeline of a semester or the walls of a classroom.

As we ended the semester, new questions and challenges emerged. What are the implications, in terms of a public educational campaign, if the farmers market patrons, who clearly are aware of the environmental dangers associated with *not* buying local food, choose overwhelmingly to frame their decision to buy local food with positive language and to leave out the fearsome or depressing possibilities from their discussion? What should we make of the omissions of race, history, access, and affordability in the responses? If people best relate to the local food movement in terms of positive associations, then the challenge for me as a rhetorician is to determine how best to use this knowledge in an educative way to help nonprofit and other local groups deliver their messages while also encouraging development of critical food literacy that connects to systemic issues.

Salsa and the "Local Trap"

Based on the three semesters of student interviews, discourse analyses, and the farmers market responses, I suspected that a community-wide food literacy campaign was needed to address the dissonances within Boulder

County's local food rhetorics and conflicting definitions of "local food." My concern is that, as Born and Purcell remind us through their term "the local trap," not all local food is ethically or sustainably produced food. Not all locally produced products would have a smaller carbon footprint than similar products produced globally. Furthermore, what is ethical to me may not be ethical to others and vice versa. My definition of sustainable production might be different from another's definition. One person's cultural heritage may be another's genocide. We cannot promote "local food" indiscriminately. In a 2014 talk given at Local Food Think Tank in Denver, local food activist Michael Brownlee said, "We were often confronted with the question, 'What's your definition of local?' . . . We quickly came to realize that this is a kind of *trick question* that gets people all tangled up in a meaningless debate. Our answer became very simple, 'As local as possible.' It's that simple" ("Increasing" 12, original italics). But Brownlee assumes a shared understanding that my research suggests cannot be presumed.

If we are only considering miles in our definition, for example, beef from the notorious factory farms and slaughterhouses in Greeley, Colorado, made famous in Eric Schlosser's *Fast Food Nation*, is local meat; flour milled in a Conagra plant in Denver is local; tuna canned in a Denver facility is local. These are extreme examples, but the definition gets murky in more complicated examples, as well. What about the Boulder-based organic bread company that sources ingredients from around the country but produces the bread in Boulder? Should that be counted as "local food?" Or the homemade salsa that used to be sold at the Boulder Farmers Market? The producer got their tomatoes from Costco. If the companies that produce the food that will be shipped nationally or globally are local to Colorado, is that food necessarily "local?" In fact, the example of the "local" Conagra flour comes from a digital presentation given by a Boulder County Parks and Open Space representative at one of the 2009 Boulder County commissioner public hearings about GMO production. Even in Rhetoric and Writing Studies scholarship, the nuances have not always been made clear. In contrast to "global food," Eileen Schell argues, "local or regional food costs less to transport, requires less preservation or modification, and is not dependent on pesticides and nonorganic fertilizers that are common in monoculture production" (91). Sometimes this is true, but pesticide-laden monocultures are local to somewhere. From the examples so far, we cannot assume that distance alone is a surrogate for all of these issues.

STUDENTS WRITE THEIR OWN DEFINITIONS FOR "LOCAL FOOD"

In my fourth semester teaching courses that delved into different parts of the local food ecology, as my students and I discussed final project options, we kept coming back to the example uncovered by my previous class of the farmers market salsa made with Costco tomatoes. Students seemed genuinely bothered by this, which in and of itself was fascinating to them. They had never purchased anything except store-bought salsa made with ingredients they'd never considered. What was it about a person selling this product at the farmers market that frustrated them? We got into the question of how *they* think "local food" should be defined and the transparency they expected and wanted.

I told them about another local salsa producer, Carmen Pacheco-Borden, founder and owner of Carmen's Salsa. She cooks small batches of salsas and molés at a time, and sources local tomatoes, onions, peppers, and garlic whenever she can but also purchases spices from purveyors across the world. Is there a difference in her product versus the other local salsa producer's product that might elicit a different response in students and customers? When I interviewed Carmen for this book, I asked her this question and whether the ingredients she sources tie to any educational, cultural, or ethical mission. Her response not only tells a beautiful story of cultural food sovereignty and education but also challenges definitions of local food that are based solely on distance traveled. Carmen explains, "I quickly realized that my mission was to tell people that Mexican food is vibrant, that it is healthy. And it's not the type of food that you find in Tex-Mex refried beans with iceberg lettuce everywhere with melted cheese. That doesn't represent the Mexican culture. It is healthy and vibrant, and it doesn't have to be spicy, and it doesn't have to be oily. So I made it my mission to have a booth [at the farmers market] that was informational and educational in addition to selling the product.

"If you visit me at the farmers market," Carmen continues, "you would often find me showcasing the ingredients, the real ingredients, at my table, especially for the molés. The first one that I learned, it's a 30-ingredient molé, and I had most of the ingredients there, the groups. And then I always explain them by groups of foods, you know, that we have nuts and seeds like pecans, almonds, and peanuts. And then you'll have the spices like cumin and cinnamon, and then you have to have the right kind of cinnamon, whole spices not ground. So then there's a difference between grinding your own spices versus using already ground spices. It was educational, and it was always with the purpose to teach them that there's different kinds of molés. The molés are not a mystery. These are the ingredients in this, and these are the ingredients that are in that. In my molés, it's a fusion cuisine that combines the old world you know, they brought the spices, you know, also with the new world. The pre-Hispanic with the new world, like you know, my ancestors with the tomatoes and chili peppers. It's a marriage of the old world with the newer.

"And so I get to talk [with customers] about archaeology and history, and it's just the marriage of cultures into a molé or salsa. And it becomes really interesting. I've had the most interesting conversations with people who walk by who have traveled in Guatemala or Mexico or different parts of the world. So to me, being at the farmers market, the mission is to teach people that Mexican food is healthy, vibrant, and also to teach them how to cook and where ingredients and recipes come from.

"I've worked with the Slow Food Boulder County, and we've held classes of making molés. So it's not about guarding the recipe because that was a recipe that I went to Oaxaca to learn. And I mean, I did pay for the classes, but it was something that I learned and then brought to Boulder. And then I went to pass it on to other people who are interested in making traditional molés from scratch, you know, the 30-ingredient molés. I share it. And you do want to have a good recipe because it's 30 different ingredients. You want to make sure each ingredient, you have to refine it. Like the tomatoes have to be roasted and peeled. You want to lightly toast the cumin like the Indians, when you cook Indian curries; they also toast the cumin a little bit to release the oils before you crush them. So, there are similarities between curries and molés. And so anyway, that's what the farmers market is for me, it's the education part. And to tell people that Mexican food, it's a refined food, and give it a chance, and it's healthy. All the molés that I produce are gluten free naturally. They're vegan. And there's no oils added at all."

I ask Carmen whether part of her educational mission is to show how critical it is to have cultural variety at the market, to not exclude groups from thinking they can see their culture's healthy food there too. She agrees that this is a key consideration and a reason why she joined the farmers market board of directors. We in Boulder County need cultural variation and representation in food conversations. Carmen's products exemplify a need for seeing complexity and having flexibility in definition building. Sometimes it is impossible to source all local ingredients. Then what are the qualities of her products that explain why the customers (and my students) more readily accept Carmen's Salsa as aligned with "local food" values? She posits an answer: "This is something that you cannot find at a grocery store. Nobody's going to take the time that I do to produce a salsa that takes about two days to make, not counting the time to have the ingredients ready when I need to cook them at the commercial kitchen. Everything is made from scratch, everything, you know, like even the spices are not pre-ground, the tomatoes are not canned. I try to buy tomatoes also from other local farmers to say, this has Rocky Mountain fresh tomatoes. When it's possible, you try to partner with farmers that have the garlic, the tomatoes, the peppers, the onions.

"And then I explain to [customers], you don't understand that the cacao, which is the raw chocolate. This one comes from a company called Ritual Chocolate, local from Denver, it's a company that specializes and sourcing high quality cacao beans from different parts of the world. I buy from them. They know me and what I do. So they recommend cacao beans to me that I should buy. So they have one from Mexico and Chiapas. I told my customers, I know where each ingredient here comes from. I source it myself.

"You're also talking about farmers and farmworkers who are not being treated fairly out there. Who is paying the price for the $2 salsa you're buying at Costco? I said, with mine, you have to understand that it is made every Friday fresh for you to consume. And Saturday it hasn't been there for months and months waiting for you. It's not acidified. And it's designed for a fresh experience. I explain how I roast the tomatoes, how I peel the chili peppers, and I give away the process, and maybe they can make it. And they realize, you know, it takes some time to get the ingredients, make sure the ingredients are ready. Like the tomatoes fresh from local farms, you don't want to make salsa when the tomatoes are not ready or when they're rotten either. I've taught my kids, and we take a lot of time on selection of the produce. And we are very careful to make sure that we only use fresh ingredients and the top quality. And then once I explained to them, all the work that went to producing the salsa, then either they don't care, which is not my customer, or they care, and they buy it. And then once they know the value and they taste it, they come back" (Pacheco-Borden).

Students were fascinated by Carmen's story and made connections to several of the values—like relationship building, knowing the producer, knowing the source of ingredients, higher-quality, healthier food—mentioned by farmers market patrons in the canvassing responses my previous class elicited. The students also made connections to the restaurateur several semesters back who asked my students to think beyond a miles-based definition for "local" and instead to think in terms of relationships between small-scale producers around the globe.

These students determined to investigate their own food selections and how they may or may not align with how each individual student defined "local food." For some, the environmental issues such as packaging and carbon footprint mattered most. Some were most concerned about how animals are treated. Some cared most about worker conditions. Others were focused on health or cost. We decided on what we called an Eat Local Project.

This ten-day final project involved collecting qualitative and quantitative data about what's really involved in eating locally. For five days, the students would eat as they normally do, keeping a log of everything they eat and drink. Based on the country or state of origin listed on packaging or the company website, they tried to determine the food miles associated with each meal, snack, and drink. By the end of the five days, they not only had a sense of how far each meal traveled and how many food miles they accrued over the block of time; they also weighed or measured all of the trash daily, saved all receipts to tally the expense of a regular five days of eating, and looked into business practices and working conditions at the food and beverage companies. Each day they wrote an analysis to accompany their log, including insights and observations, as well as ideas and questions that surfaced as they went through the process.

Over the weekend ending the first five days, with funds from a grant I'd received, I shopped with students to purchase local food for them for the next five days of the project. Most of them met me at the Saturday farmers market a few blocks from campus to purchase five days' worth of local food to eat during the following five-day period. If they needed additional food or beverages, we went to a nearby Boulder grocery store that labeled local food items. Students recorded each farm or company, tried to determine how many miles the food traveled, and completed other qualitative and quantitative assessments of the food. They turned in the ten days of log and daily analysis along with a longer critical analysis of outcomes and learnings from the project.

Through this exercise, they noted how difficult it was to track food origin for all of the ingredients in industrial food. But for the grocery store purchases, they noted that some items marked "local" are processed locally but with ingredients from elsewhere (much like Carmen's Salsa). Examples are a local coffee roaster that sources beans from various countries that the owner visits himself to establish relationships with small growers and know growing practices firsthand, and a local pasta company that uses wheat shipped from Italy labeled as "Local." Others noted that they could eat well for the full five days on food grown entirely in Boulder County. Some students noted that the grocery store we visited had a choice between local meat from a factory farm or pastured meat from another state—here they had to choose between distance from which meat was shipped and animal welfare in terms of which mattered most to them. They argued, however, that they needed to account too for not only distance the meat traveled but also the animal feed shipped to the factory farm used to feed the local cattle. They realized that much more goes into how food miles are calculated than simply end product to plate. Others realized they could buy less expensive items like quinoa and dried beans in bulk at the farmers market to sustain them through the week, and that this was less expensive than an equivalent purchase at a store. Many of them noted the energy and friendliness of the farmers market compared to the grocery stores, perhaps because they needed to engage with people and ask the farmers questions about the farm location and practices. Some students found it difficult to grocery shop for local foods that were also culturally relevant. Some chose to pay a bit more for products from two farms we had visited earlier in the semester because they felt a connection to the farmers.

These are only a few of dozens of issues students considered, experiencing firsthand the complexities involved in their deliberations and the food literacies required as they defined what mattered most to themselves in their food choices as they navigated the various systemic issues entangled in the typically mundane task of choosing what to eat for a meal. Assignments like this one can bring to life the problems associated with imposing definitions and can deepen understanding of nuance. It is easy for students to read about issues and create simplistic binaries—eating in one way is good and another way is bad. This assignment manifested the messiness we'd been discussing all semester and allowed them to see how important the previous classes of students' research was in helping them develop their own food literacies.

STUDENTS WANT THEIR WORK TO MATTER

My class of students who had rented the farmers market booth concluded that given how complex and subjective a definition of "local food" is and the many different considerations that patrons mentioned in the responses they received, no one person or agency or organization should have the sole power to define it. Instead, we agreed as we broke bread together around my dining room table, local food taps into profound human and nature-based connections that people crave (often without even understanding the systemic reasons for the disconnections that are ubiquitous throughout American culture in the United States). Each community member should be able to build that desire for connection into their own definition based on what is most important to them. The ability to build definitions should be distributed across community members.

I committed to my students that night that future classes would not only read their and previous students' findings but would work with community members directly in a distributed and collaborative food literacy process that replaces top-down and sometimes uncritical definitions with ones that grow from the community itself. This commitment helped me to name what my students and I had been doing for several semesters. We were studying and helping to create (sometimes linguistic, sometimes physical) connections between various groups, media, and communities in our local food ecology. In creating those connections through multiple approaches that solicited a wide range of definitions, we were developing a relational methodology of *ecological community writing* through *distributed definition building*.

3

Crisis and Abundance Rhetorics in Moving People Toward Collaborative Engagement

> *The challenges of the coming years look like a house of cards—one disaster collapsing into another. They feel impossible to map—or grasp, even—in their immensity and interdependencies. Those same linkages, however, are potential nodes of connection. Our fates are bound, awaiting the story that will bring us together.*
> —V. JO HSU, *CONSTELLATING HOME*

Rhetoric and Writing Studies has a many-decades-long history of scholarship addressing connections between environmental issues and ways in which they are written about and conveyed to various groups of people from students to international audiences (Hawhee; Opel and Sackey; Owens; Rice; Sackey and DeVoss; Walker). We grapple with how to make various publics care about multiple crises for what Derek Owens calls "a threatened generation." Community writing practitioners have a delicate balance to strike between imparting the complexities of systemic issues that could cause people to withdraw or shut down and encouraging community members' participation in conversations to keep them engaged.

For example, developing narratives about community food literacy can walk a line between depressing and exciting, traumatic and joyful. As someone organizing food literacy projects and events, I need to give information and create opportunities necessary for people to make more informed choices and dig into why they make the choices they do or feel the emotions they do as they make those choices. At the same time, I need to present the information in a way that avoids overwhelming people to the point of emotional shutdown. The information needs to be accessible and engaging in ways that root into profound feelings of connection that so many humans crave. Community writers and teachers need methods for striking this balance so that people will engage. This chapter delves into some of the research from multiple fields of study that has impacted my praxis.

Most semesters, I see my students hit an emotional wall about a month into the semester once they've completed their research and the readings on industrial food production. Inevitably they get to a point, regardless of their dietary preferences, where they're wondering whether there is anything they can eat that doesn't exacerbate several environmental and social problems. We usually have several conversations and do some reflective writing about what they are feeling, and there are key ideas that I foreground. First, not to fear, we will find meaningful ways to engage with food that align with their values, however they might individually define those values. We'll bring in local experts to tell us about innovative local projects, and we'll get to see these projects firsthand through work visits to sites around the county. Second, because students often ask me how I and my family eat, I assure them that I understand how complicated it is, that I also struggle with all the choices we can consider, and that perfection is not the goal. Even small decisions matter, and we're all balancing a variety of legitimate factors from health to convenience to financial means to ethics as we make our daily choices. As with all discussions in the classes, I try to encourage an environment in which we're all asking and exploring together without "correct" answers and imposed definitions.

Even as we turn to our study of local food, the numbers and hugeness of the issues can leave students feeling overwhelmed. This overwhelm can happen with community members too. When students and other community members realize, for example, that definitions of local food in constituencies across Boulder County are so varied and subjective, they could very well determine that literate decision making is a lost cause or that their personally determined definitions don't matter. As my students and I discuss the overwhelm we feel, study definitions from multiple parts of the local food ecology, and consider other places in the ecology where outreach and education could occur, we ask ourselves how to communicate effectively about complex and sometimes traumatic issues while maintaining excitement. It's not an easy thing to do. In thinking through how to create that kind of environment in our classrooms and in our communities too, rhetoricians can encourage action through strategic creation of language and projects that resonate with the positive effects of engagement while maintaining focus on the seriousness of issues. This tension led me to develop a concept of *critical rhetorics of abundance* that I use in the community food literacy work I do.

Climate communication studies and psychological studies on trauma have helped me understand how to motivate action in students and community members. As I will explain in this chapter, using *critical rhetorics of abundance* is a way to frame *distributed definition building* projects and events with students and community members across multiple positionalities, ages, and levels of engagement. Unlike industrial food rhetorics of extraction, efficiency, and excess, *critical rhetorics of abundance* embrace relationships, connections, emotional engagement, and culturally meaningful foodways. They embrace a complexity of emotions that allows for trauma and pain and anger to be acknowledged alongside joy, celebration, and gratitude while moving people toward justice, accountability, and transformation. When framed through language and experiences that embody *critical rhetorics of abundance, distributed definition building* offers a method for community writing practitioners to engage communities and classrooms in interrogating complex and systemic issues so that people stay invested.

Why People Shut Down

Local food resilience is both an environmental issue and a social justice issue tied explicitly to many of the topics I discussed in chapter 1. Environmental and food issues more broadly cannot be divorced from corporate and racial capitalism, fossil fuel dependence, hunger, settler colonialism, and more. Any major disruption to national or global supply could lead to a food crisis. Boulder food activist and writer Michael Brownlee reports that in 2006, a working group in Boulder County calculated that if food imports were cut off, "with the current food and agriculture system, we could feed only about 20,000 people" in the county on an ongoing basis (Brownlee, "Local Food and Farming"), while the county's population at the writing of this chapter is upwards of 322,000 people. A 2015 accounting from then–Boulder County Farmers Markets executive director Brian Coppom indicates that "the entire season's sales for the Boulder and Longmont, CO [farmers markets] could feed [those 322,000 people] for only . . . a day and a half" ("Interview"

81). When faced with these numbers, it becomes clear why so many Boulder restaurateurs and chefs struggle to attain local food. The stark divide between local food production capacity and population is not unique to Boulder County. Some food localization activists in Boulder County and across the globe have tried to force into the public's consciousness difficult questions regarding who may and may not have access to food in times of infrastructure breakdown or climate-related food shortages, and answers are inevitably tied to wealth and racial breakdowns of communities.

The implications of the numbers are grim. If only 20,000 of the 322,000 Boulder County residents would be able to eat consistently if there were long-term disruption, what would happen to the other 302,000 people in the county? Crisis? Hunger? Famine? The words seem apt if a sustained, severe disruption were to occur. And yet, the possible answers are so disturbing, I can't fully wrap my mind around them. When I, as a local food advocate, receive emails like the one I cite here from a local food activist group that arrived in my inbox while writing this chapter, my initial, powerful impulse is to quickly press delete: "Industrial civilization is rapidly heading towards collapse, and our biosphere is experiencing its most disastrous crisis in the last 65 million years." I feel my body tense up. My heart rate speeds. I believe what the email says, and I care deeply, but I don't want to see it. Though of course I cannot, I want to push it away. It is too tragic to know how to process. In her study of climate crisis rhetorics, Debra Hawhee cites several young people who testified in 2019 before the House Select Committee on the Climate Crisis. They attest to feelings of betrayal coupled with an extreme sense of urgency and with severe anxiety and depression. As one youth testified, "I have felt as if there is a pressure cooker boiling over inside of me. I can hardly focus at times because I am overwhelmed with existential horror at the fate of the planet" (qtd. in Hawhee 65). My students express similar reactions to crisis rhetorics.

In *ecological community writing* work with multiple groups across an ecology, rhetoricians and writing studies practitioners can help people to create language and stories that avoid reinforcing crisis frames of deprivation and doom that inadvertently may lead people to reject information, shut down, or despair. As I've detailed in previous chapters, many argue that we are living through a food crisis that will worsen with climate change. Rhetorics abound of shortages, disease, famine, and corporate control of food and water, as well as apocalyptic rhetoric about the uncertain future of food availability given global climate shifts. Add in topics like soil depletion, contamination of our waterways, modern-day slavery and abuse, microplastics and heavy metals found in fish, and on and on, and the amount of disturbing information can become overwhelming. In addition, as Hawhee has argued about people's inability to grasp the concept of exponential growth, the magnitude of these concepts is "difficult for human minds to grasp in the abstract—[they are concepts] that lie beyond most humans' intuition" (104).

Drawing on the positive frames used in the local media sources and farmers market responses that I discuss in chapter 2 and on significant research in climate communication and trauma studies, I've adopted a lens of *critical abundance* through which to approach *ecological community writing* to help shift framing of an issue to move people from overwhelm and potential inaction to engagement in local community efforts. I now approach projects with students and community members through *critical rhetorics of abundance* rather than crisis or scarcity rhetorics. As many rhetoricians have argued before me, the framing of an issue matters.

In this chapter, I turn to climate communicators and psychologists who have guided my thinking about how to encourage affective engagement and action, and whose conclusions influence my conception of critical abundance that informs how *distributed definition building* projects and events can be framed. I look to these scholars and studies not only because of the interconnectedness of climate change, agriculture, white supremacy culture, and trauma but also because there has been important research on what kinds of messaging about traumatic issues make people shut down and refuse to act versus open themselves toward engagement. Climate communication and ecopsychology research are useful to consider in *ecological community writing*, which involves creating accessible content to communicate about social, cultural, or environmental issues in our classes, community work, and research. I hope that in drawing on these various disciplines for ideas of how to engage community members most effectively in possibly difficult conversations, I can add to the rich Rhetoric and Writing Studies scholarship on climate and environmental rhetorics happening in the field (Hawhee; Rice; Sackey; Walker). I will then turn to how *critical rhetorics of abundance* offer an effective lens through which to translate the work in the other disciplines into an *ecological community writing* methodology. Finally, I will introduce a community partner, The Shed: Boulder County Foodshed, with which I worked for several years on multiple food literacy projects across the county.

The Problem of Rhetorical Overwhelm: Lessons from Climate Communication, Ecopsychology, and Trauma Studies

In my work with students and with community food literacy projects, I have been influenced by several studies that analyze the effects of different kinds of messaging on people's psychology and behavior. Much of the rhetoric around food campaigns has historically drawn on the moral panic of informed eaters (Hahn and Bruner 50)—a fear of agribusiness, fear of foodborne illnesses, fear of environmental collapse, fear of corporate capitalism, and so forth. This consumer anxiety has been used to attempt to generate change. But, as Jenny Rice warns in her study of a "nonparticipating subject," "feeling easily becomes a substitute for action. That is, feeling becomes both the evidence of and the actual activity of public relationality. Feeling angry [or fearful] is not a prelude to action; it is the action itself" (60). Apathy too, she posits, is "a political response to the fragmented and non-representative character of the public sphere" (66). She explains, "[C]itizen nonparticipation [is] an effect of certain rhetorical patterns within current public discourse rather than a symptom of disengagement or misinformation" (6). Given the need for a local food literacy campaign in Boulder County, I needed to imagine a model that would encourage emotional engagement and participation, feeling *in* action, moving beyond that emotional wall people may hit. But how?

The rhetorical discourses that activists often use in relation to food, like the quote cited earlier from the email I received from a local food activist, may inadvertently encourage nonparticipation for a number of reasons. It can be because they are crisis based, because there is such confusion over what is and is not safe to eat, what is and is not ethical, and what the eater is giving up to eat a certain way, and because of deliberate obfuscation on the part of industry, as I discussed in chapter 1. All of the competing discourses and definitions related to our food production system and dietary choices can lead to feelings of anxiety and frustration that end in a disillusioned decision not to act. I imagine that this internal struggle may sound something like "What *can* I eat? There are too many considerations. It's too confusing. I don't want to deal with this. I'll just eat what I want."

Some psychologists and evolutionary biologists argue that humans are not wired to tolerate well emotions such as fear and anxiety over sustained periods of time (M. King). Perhaps that is part of why so many millions if not billions of people refuse to act decisively on the overwhelming evidence presented by climate scientists. There are cultural-conditioning and spiritual practices as well that play into people's willingness to turn from or toward suffering and trauma. It is worth looking at some of the recent studies in order to apply them to framing of the local food movement and other food literacy campaigns. Because I draw from disciplines outside of Rhetoric and Writing Studies, I will quote and summarize at some length as readers may not be familiar with the scholarship.

PSYCHOLOGICAL RESPONSES TO CLIMATE COMMUNICATION

Over the last fifteen years, psychologists have witnessed a sharp spike in environmental anxieties, so much so that a subfield of Ecopsychology has developed. In fact, worldwide, "psychological researchers have documented a long list of mental health consequences of climate change: trauma, shock, stress, anxiety, depression, complicated grief, strains on social relationships, substance abuse, sense of hopelessness, fatalism, resignation," and so forth, "because of our deep sense of fear and anxiety [that] underlie our concern for the future" (J. Richardson 174; Hawhee). Anxiety, anger, panic, and depression are common in climate scientists as well, stemming from what psychiatrist Dr. Lise Van Susteren calls "pretraumatic" stress (qtd. in J. Richardson). The average person cannot cope with the kind of chronic stress that climate crisis discourse can cause, so the alarming scientific discourse may inadvertently *cause* inaction. These symptoms often manifest as fear, anxiety, paralysis, and denial (Lertzman 5). These are defense mechanisms that we often cannot even consciously explain.

I'd like to turn to findings from recent studies and scholarship about climate communication and ecopsychology that address these issues, because, to my knowledge, they have not previously been written about in community writing scholarship. The findings proved vital in the development of my *distributed definition building* method onto which I incorporate a *critical rhetorics of abundance* lens for community engagement projects. The findings also prove useful pedagogically to encourage student engagement. They serve as a reminder for how my project partners, students, and I might create more effective communication around the local food movement in Boulder County, especially given the definitional complexities and confusion I discuss in the previous chapters.

There are two strands of response that happen simultaneously when humans hear about a social issue—the socially motivated and the individual—and I will address each, as they impact how rhetoricians and writers can communicate and teach about the issues. It is no revelation that humans are social animals and that we are immensely impacted by groupthink and by the cultural milieu in which we live—our family, friends, peer groups, co-workers, and so on. Climate change communication specialist George Marshall argues in his book *Don't Even Think About It: Why Our Brains Are Wired to Ignore Climate Change* that "the reason why people do not accept climate change is nothing to do with the information—it is the cultural coding that it contains" (23). This is partially why no matter how many facts scientists offer, some people continue to refuse to believe in climate change. It is not a matter of presenting more or better facts; rather, it is critical to understand the latent issues driving behavior. Marshall explains, "[R]ational scientific data can lose against a compelling emotional story that speaks to people's core values. . . . [C]ommunications from people's family, friends, and those they regard as being like themselves (their peers) can have far more influence on their views than the warnings of experts" (24).

The lack of success some climate scientists have had to effectively move people to act reveals the powerful human desire to conform. Marshall explains, "When we [humans] become aware of an issue, we scan the people around us for social cues to guide our own response: looking for evidence of what they do, what they say, and conversely, what they do *not* do and do *not* say" (27). This is wired into human behavior as an evolutionary development to aid in our survival as part of a group. As cognitive linguist George Lakoff explains in *Don't Think of an Elephant*, this means that it is critical that rhetoricians and writers think through the "we" in public messaging. If communication does not resonate with a specific group's values and emotional story, it can often inadvertently repel rather than attract them. A clear recent example is the diametrically opposed responses to Covid-19 mask wearing and vaccines based on people's social groups and the anxieties and fears those groups trigger.

People apply their existing "frames," to use Lakoff's term for the "mental structures that shape the way we see the world . . . structures in our brains that we cannot consciously assess" (xii), to an issue to see whether it resonates or not within their worldview. Drawing on Lakoff, Marshall explains, "Frames are like the viewfinder of a camera, and when we decide what to focus on, we are also deciding what to exclude from the image we collect" (80). Marshall calls this *"disattention,"* "when we deliberately fail to notice something and cannot even explain that silence" (82). A similar concept is *confirmation bias*, "the tendency to actively 'cherry-pick' the evidence that can support our existing knowledge, attitudes, and beliefs. These create a mental map—what psychologists would call a schema—and when we encounter new information, we modify it to squeeze into this existing schema, a process psychologists call *biased assimilation*" (14, original italics). When applying these concepts to climate change deniers, Lakoff explains, "Many conservatives don't oppose climate science because they are ignorant. Rather, it is a way of expressing who they are. The obstacle becomes the innermost barrier to climate communications: The messages crash against the wall of the self" (74).

Lakoff's revelation is that neural circuitry in the brain holds our morals because each time a belief is reinforced, the neural pathways in the brain trigger. This means that "we may be presented with facts, but for us to make sense of them, they have to fit what is already in the synapses of the brain" (16). In other words, there are neurological reasons for disattention, confirmation bias, and our propensity toward group think. "All thought is physical," Lakoff explains, "carried out by the neural circuitry in one's brain. Thoughts don't just float in air. As a result, you can only understand what your existing brain circuitry allows you to understand" (51–52). Thus, Lakoff claims, "moral sense, like all that we believe and understand, is physical, built into the neural circuitry of our brains ... Our political divisions come down to moral divisions, characterized in our brains by very different brain circuitry" (41). This means that when we rhetoricians, writing teachers, organizers, and others address an audience about community food literacy or another complex issue, we need to consider what people's brains may be receptive to hear.

These findings can help shape the way we work with students on any issue important to our communities. A useful exercise could be to generate a list of beliefs or areas of connection of the people or groups the students are engaging and then to generate rhetorical frames that might make sense to that person or group. A plurality of stories that can reach a wide number of audiences and address an array of belief systems is needed (Hayhoe and Farley). If climate communication has focused on disaster, apocalypse, sacrifice, death, and destruction, we can shift the framing to be about joy, resilience, cultural sovereignty, imagination, freedom, health, community, and connection.

To take these concepts a step further, there are two kinds of causation, Lakoff teaches—direct and systemic. Direct is easy to see: You touch a hot stove, and you get burned. On the other hand, because systemic causation is not linear or direct, it can be literally impossible for a person to comprehend if their brain is not used to thinking systemically. But studies of neuroplasticity indicate that neural pathways can gradually be re-formed. Effective communication regarding systemic crises may require an understanding of a person's baseline beliefs so that they can gradually be moved in a different direction. The organization generative somatics, which supports "social and climate justice movements in achieving their visions of a radically transformed society," encourages us to "engage the body (emotions, sensations, physiology) in order to align our actions with values and vision and heal from the impacts of trauma and oppression." From work on contemplative practice in Writing Studies to somatic work in psychology and social justice organizations to scientific understanding of the mind-gut connection, multiple fronts are working with understanding the physical nature of our morals, traumas, fears, learning patterns, and so on, which are not always conscious.

In addition to the cultural and social systems in which we live and process information about threats, individual emotional and affective processes occur that prevent people from being able to accept the enormous implications of climate science. Jeffrey Kiehl, who was a senior scientist with the National Center for Atmospheric Research in Boulder, Colorado, prior to earning a master's degree in clinical psychology, explains the central question that drove him to return to graduate school: Why do humans turn away from factual messages of the overwhelming evidence of climate change? Similarly, Per Stoknes wonders what is behind this "massive exhibit of the human capacity for self-destruction" (25). He calls scientists' inability to reach people in effective ways that spur action "the greatest science communication failure in history: The more facts, the less concern" (81). Humans often need "felt evidence—which differs from a calculated, deliberative, rational process of 'weighing the evidence"' (Hawhee 111). Across disciplines and communities, environmental communication scholars grapple with these questions of how to reach a wide range of unreceptive audiences.

In *Facing Climate Change*, Kiehl explains that he found an answer to these questions in social psychology, which defines a type of person called a "security seeker": "[T]hose who place great value on personal security ... If the security seekers experience fear, their instinctual reaction is to protect themselves against whomever they see as a threat. They resist change even in the face of factual evidence arguing for change" (21). What's more, this is not only culturally constructed but an evolutionary survival strategy: "Our way of being is defined by patterns of beliefs and behaviors and may be so ingrained in us that the thought of altering them fills us with anxiety, even though those patterns are destructive" (17). Climate change facts are extremely threatening to the ideologies of consumerism and individualism that so many people living in the United States identify with as core cultural values. The emotions of "sadness, hopelessness, anger, denial, guilt, numbness, and fear" provoked by climate communications crash up against a desire for quantifiable and individualistic abundance, for a certain lifestyle, or for a perceived sense of freedom and agency (29). Through his master's program, Kiehl realized he needed to stop presenting scientific facts alone, instead "allowing time for people to express their feelings about those facts. . . . [All of those emotions listed above] are indicators of trauma" (29). "What comes to the fore," he explains, "when I listen to people's emotional responses is the tremendous sense of felt and perceived loss in their personal worlds" (29). Indeed, people's feelings of loss surface in all the studies I discuss in this chapter.

Psychological studies explain that when people feel threatened, they adopt several (often unconscious) strategies to diminish feelings of loss and anxiety. Renee Lertzman, a psychologist who focuses on climate anxiety in her book *Environmental Melancholia*, expresses an idea similar idea to Kiehl's: that loss dominates the responses people have toward environmental threats. Her study uncovers "the central role of unconscious processes, specifically guilt, anxiety, loss, and related defense mechanisms in how we relate with our ecological threats" (31). In her interviews with people who do not consciously identify as environmentalists, narratives of loss dominated. Interviewees described loss of place or of environmental objects such as a beloved lake or river. Often these places were connected to childhood and innocence and to a place of "refuge and sanctuary" (84). Thus, Lertzman argues, "the external 'presentation' of apathy, appearing as an *absence* of action, may in fact be related to loss" (101, original italics).

Marshall's theory corroborates Lertzman's study, centering on what he calls "the Great Grief, a feeling rising in us as if from the earth itself at this time" (171). "The grief was not triggered only by climate change itself," he explains, "but by the extended loss of beauty, diversity, and habitat" (172). Marshall posits, "We need to **MOURN WHAT IS LOST**. And not just the natural world; we need to **MOURN THE END OF THE FOSSIL FUELS AGE**, which, for all of its dirt and danger, was also exceptionally affluent, mobile, and exciting" (238, original caps and bold). It should be noted that Marshall's "we" is not universal or monolithic. Also worth noting, Lertzman's findings on grief and apathy may present differently in different cultures. In some ways, however, virtually all people who live in the United States have become inextricably reliant on fossil fuels and products and foods available only because of vast fossil fuel usage. This makes the mourning Marshall suggests extremely complicated.

In *The Inner Work of Racial Justice*, Rhonda V. Magee explains spaces that feel white as "shaped by a culture that has privileged the experiences of white people. It can be characterized by hyper-individualism over collectivism, entitlement to material comfort if not luxury in a world of presumed scarcity, and a sense of disconnectedness from the broader context in which we live" (ch. 15, 3:08). While Magee is talking about a physical building or room, the idea of white space can apply to the United States as a whole, and the feeling Magee describes echoes in United States mainstream culture writ large. The fossil fuel age and what it has afforded are bound up in the white capitalist ideals Magee mentions. adrienne maree brown writes about her own grappling with this complicated entanglement, "It is perpetually disgusting to contend with the reality that these disasters benefit a bloated elite. And too many of us participate in our small-scale versions of their individualistic and hoarding worldview, thinking we are better than each other and the earth, deserve unlimited resources and access, and should never have to adapt to protect others" ("Darwin"). From things like plastics and cars and cheap industrial food to systems like white supremacy culture, many people in the United States believe they benefit from the very things climate scientists tell people to give up, which leads to confusion, frustration, resentment, denial, and the other defensive emotions.

Psychologists suggest that humans cling to the things we fear we'll lose. When there is personal guilt associated with loss, Stoknes explains, "Our creative minds quickly find a solution to this sorry state: *Rather than change my actual behavior, I can modify my thinking to match what I do*. In social psychology, this way of dealing with our attitudes has a special name: cognitive dissonance" (61, original italics). These feelings, Lertzman posits, lead to "environmental melancholia," "a condition in which even those who care deeply about the well-being of ecosystems and future generations are paralyzed to translate such concern into action" (4). She explains that actually "most people care very deeply about the quality of life on the planet and the desire for future generations to enjoy a vital world." However, "because a deep sense of fear and anxiety underlie our concern for the future," people may experience "paralysis, defensive mechanisms such as denial, projections, splitting, and dissociation" (5). These findings are particularly important in that, based on people's inaction, climate communicators have made possibly incorrect assumptions about people not caring about climate issues or not understanding the facts about climate issues. So, they may pile on more facts and explain the situation as increasingly dire, thinking that these strategies will make people care more. Lertzman says that the opposite is often true. Rather, people care so much that they generate often unconscious responses to push away the issue.

A critical element of people's responses to climate change is the perceived lack of its effects' proximity to them and their daily lives. People are generally able to respond to immediate crises, but it is much harder to respond to an anticipated, future crisis. If we are not part of the world population seeing the direct impacts every day, we can push it aside. Boulder writer and social entrepreneur Matthew King explains that there are evolutionary reasons for this: "Throughout most of our evolution it was more advantageous to focus on what might kill us or eat us now, not later. This bias now impedes our ability to take action to address more distant-feeling, slower and complex challenges." It is challenging to ask people to give up what is familiar and comfortable for unseen future challenges that are very difficult to imagine. If people who do not yet live in highly affected areas respond to the enormity of what scientists say we face with apathy, disassociation, and other responses mentioned by ecopsychologists, they may cling even more fervently to their current way of life. In Boulder County, where climate change is having a direct impact on local agriculture, and farmers are concerned for their livelihoods, difficult conversations with them and the public are ongoing and often difficult.

This is true for many parts of the world, including parts of the United States, currently seeing the effects of climate change through hurricanes, floods, fires, toxic waters and soils, and oil pipelines, and the resulting disruptions to health care, food, and other basic needs. According to Dr. Annelle B. Primm, those hardest hit by current and future effects of climate change are often communities of people of color and disabled individuals, those who live in high-risk areas due to environmental racism, ableism, fewer resources, and unemployment. Primm says that well-being is rooted in "presence, connection, and interdisciplinary and collaborative approaches." Basic needs must be met on an individual level, but we need to think in terms of communities as well—the resources they need and, from an assets-based approach, what they *have*. In local food considerations, an assets-based, culturally specific, and geographically specific approach will impact how people define what is important and feasible for them and their community.

Ecological Community Writing, Affect, and Historical Trauma as Parts of an Ecology

As trauma psychology teaches, it is valuable to consider what individual and ancestral traumas and memories bodies carry, as they may impact the messages we convey and how we make them. The climate communication and ecopsychology work mentioned in the preceding text—some of which does not delve into race, culture, and other intersectional relationships that affect responses to messaging—can connect to (and would benefit by connecting) racial trauma psychology and somatics (practices using the mind-body connection such as meditation, yoga, dance, song, and chant). These confirm the physical nature of thought and feeling and explain the potential danger in viewing response to societal issues in terms of individual response that can be swayed through logical arguments. As significant research on intergenerational trauma and somatics shows, our reactions to trauma are not only based on events in our own lives but also can be passed down through generations in our very DNA and held deeply in our bodies (e.g., Menakem; B. Wade; Wolynn). In conversation with this research on trauma and somatics, Resmaa Menakem, a psychotherapist who studies trauma's relationship to white body supremacy and racism, explains in *My Grandmother's Hands* the racial implications of the physical nature of feeling: "Most of us think of trauma as something that occurs in an individual body, like a toothache or a broken arm. But trauma also can spread between bodies, like a contagious disease." The traumas, embedded in our bodies, can be passed generation to generation, now called "intergenerational trauma" or "historical trauma." The compounded trauma (including the inflicting of trauma on others) can lead to patterns of hatred, anxiety, denial, and pain. As Menakem reports,

> Recent work in genetics has revealed that trauma can change the genetic expression of the DNA in our cells, and these changes can be passed from parent to child. And it gets weirder. We now have evidence that memories connected to painful events also get passed down from parent to child and to that child's child. What's more, these experiences appear to be held, passed on, and inherited in the body not just in the thinking brain.

Our brains and bodies hold these complex (sometimes intergenerational) emotions.

Because emotions are held not only in the brain but in the body as well, many people are not even consciously aware of their biases and the reasons for why they react in certain ways to certain information—from what it may mean and feel like for a Black person to farm land to what it may mean and feel like for a Native American person to cultivate traditional plants or return to traditional ways of hunting. Ancestral trauma and grief as well as joy and celebration are critical aspects to consider in *ecological community writing*, which connects parts of an ecology that exist both in the present and back in time to felt and historical elements that make the ecology what it is today.

Trauma psychology and generative somatics, including contemplative writing practices, are useful in bringing people back into their bodies to sit with complex emotions as they form definitions for what matters to them in relation to social and environmental issues. Angel Acosta, who works at the intersection of social justice and mindfulness, explains how significant it is to "return to the body, to reconnect to the body." The psychological work on trauma can help to explain not only individual but group behavior. Grief expert Breeshia Wade explains in *Grieving While Black*:

> The greater the loss the more time we need to integrate it into our subjective experience. Our social location determines our access to time and thus to our ability to grieve and reintegrate after loss. It also determines our access to others onto whom we can push our grief. When we fail to recognize the grief inherent in so many of our other emotions, we fail to engage

honestly with our experience. . . . Neglecting that grief leads to spiritual decay, a rotting that attempts to plug its own holes with the wholeness of others. Too many of us are driven by grief rather than informed by it, allowing our grief of impermanence and powerlessness to justify our brutish disempowerment of others. In this way, systems of injustice are built upon grief . . . When we are driven by our grief, we enact harmful power dynamics against others in order to avoid the reality of impermanence.

The grief is very different for people of different racial and cultural backgrounds and experiences.

As a white person of European ancestry living in the United States, I need to face the ways in which ancestral violences and traumas, both those done by them and done to them, live in my body. And I need to consider this in how I define and understand healthy, local food. Though supposed benefits come from power, I need to recognize and grieve the fact that some of what I've been trained to consider as benefits to me through capitalist and white supremacist ideological indoctrination actually undermines my own humanity and connection to others and the earth. Citing the Zen Buddhist chant of purification, Breeshia Wade says that confronting loss and trauma "calls us to not only look at the circumstantial evidence of benefit and harm caused in our own lifetime but to consider what we have inherited across generations as a result of our ancestors' actions and take responsibility for how their karma has impacted our lives." Wade's comment rings true for my own confrontation with my ancestral past on my mother's side, as I discuss in "On Lineage and Purpose." If I do not delve deeply into the trauma associated with the harm I cause and that my ancestors have caused, I will pass it on to future generations. If I do not turn toward the grief, I will suppress and deny it. Wade says, "Often when well-meaning people are confronted with the abuse of privilege, they become defensive because the reality of their actions challenges the person they've imagined themselves to be." I can work, in a lifelong, ongoing process, to feel comfortable with discomfort and grief, anger and loss.

The findings about human capacity to repress feelings of trauma, fear, and threat lead to significant challenges for climate change communicators who want to both inform and encourage action. There are clear connections with community writing work focusing on any number of social and environmental issues. As I said in opening this chapter, there is a careful balance we as educators, activists, organizers, scholars, and students need to find between overwhelming people with the reality of the current converging crises humans and other beings face and allowing for complexity of emotions that include grief and joy. Community-engaged practitioners can use these findings to help guide our own work, whether through pedagogical engagement and presentation of information to students or through participation in community writing projects. Thinking back to the email I mentioned early on in this chapter, maybe I wanted to hit delete because I didn't see in it an emotional container to hold the feelings that arose; I didn't see how to act in a meaningful way; and I didn't see a space to come together with others. Would I have engaged differently if those elements were offered, if the message had instead been framed through *critical rhetorics of abundance*? I believe so.

What Can Be Done?

In the face of evolutionary, physical, and socially conditioned defenses against potentially traumatizing information, the authors mentioned earlier repeatedly suggest that the only way to experience real individual and societal transformation is to allow people to acknowledge and feel trauma, grief, and fear. Lack of engagement with the topic of climate change or any other social, environmental, or ancestral issues may represent not only a person's inability to process the loss but also "a failure to find an adequate 'home' for contributing and participating *creatively* with ecologically reparative practices" (Lertzman 139). To work beyond defense mechanisms, Lertzman argues, "we can begin to appreciate that where there is presumed to be a lack of affect, in fact there may be a surprising surplus of affect to explore" (101). This is where community writing practitioners can be particularly useful—in creating spaces for people (including our students) to find creative ways to contribute and to explore their feelings about difficult topics. Across all of the climate communication, psychology, and studies on trauma and grief cited here, one conclusion is clear—those of us focused on communication and rhetorical framing of complex issues must provide opportunities for people to acknowledge and *feel* loss and trauma profoundly and, critically, *to move into those feelings through action*. But, to bring us back to the question framing this chapter, how can we move into profound loss and grief without becoming overwhelmed and immobilized and while also feeling joy and hope?

Marshall reminds us that trauma and grief over environmental collapse are similar to other kinds of trauma and grief, and we, therefore, need to allow ourselves to move through the stages to acceptance. He explains, though, that he does not mean passive acceptance. Rather, "in acceptance we can also find resilience. The bland, bright optimism recedes, a darker, tougher version of hope is rebuilt, and new existential strategies are developed" (185). In his new version of hope, he says, "we may find a renewed way of caring for the land, air, ourselves, and others. We may shift gradually from helpless depression to heartfelt appreciation and re-engagement" (186). He calls for hope that is "grounded in our being, in our character and calling, not in some expected outcome. The future is fundamentally uncertain and complex. Therefore, it is open to the imagination and always possible to influence in some way. . . . The active skeptic gives up the attachment to optimistic hope and simply does what seems called for" (222). Marshall's last point is critical. Rhetoric and Writing practitioners can frame issues so that people feel that they can contribute *to what is called for* through collaboration, creativity, authentic participation, and imagination all tied to humans' evolutionary drive for group connection.

The authors discussed here believe in the power of compelling and resonant stories. Communications theorist Walter Fisher has argued that "when non-experts make sense of complex technical issues, they make their decisions based on the quality of the *story* . . . rather than the quality of the *information* it contains" (qtd. in Marshall 106, original italics). Again, it is not a matter of feeding people more facts. The Center for Story-Based Strategy, a national organization dedicated to using stories for social change, suggests that a compelling story must have simplicity of cause and effect, a focus on individuals or distinctly defined groups, and a positive outcome. "Positive" should not mean simplistic and naive. I would frame it more as complexity that allows joy and connection to emerge. As communicators, teachers, community members, and scholars create stories, we can be careful not to reenforce frames of doom and gloom that inadvertently lead people to ignore and suppress information. People's willingness to act may depend on so many components—from emotions to cultural conditioning to the framing we use.

I return here to the concept of abundance and the complexity required of a critical abundance. Scholars from Carmen Kynard (e.g., "All I Need" and "Oh No") to adrienne maree brown (e.g., "Radical") have argued for hope, love, and imagination as verbs, actions, and processes and practices that can exist simultaneously with anger and grief. So here *critical abundance* is an action, a practice. It rejects naive optimism and toxic, culturally enforced positivity and calls for us to grow our capacity to sit with grief and trauma, not as feelings to get over but rather to embrace and reflect on in concert with real joy and pleasure.

As more Rhetoric and Writing Studies scholars turn toward contemplative and somatic practices in classes (e.g., Mathieu and Muir), it is useful to consider how this work might merge with community writing work (House and Briggs). Angel Acosta explains how important contemplative work can be in this social justice and community-focused work, and that it has "a physiological effect on the body, has a very visible neurological manifestation, changing neuropathways in the brain." Acosta explores how contemplative science can be used to impact social behavior as well, in things like collaboration and compassion, and how contemplative practice can address "the structures that cause suffering in the first place." Contemplative work, or what Rhonda Magee calls "revolutionary mindfulness," is not just for individual well-being but also for collective justice. "Cultivating an awareness of our grief takes significant courage," Breeshia Wade says. "We must view grief and the vulnerability that stems from it as a gift, not a nuisance, because of the love and healing it can bring."

Wade's idea here is no small task. Viewing vulnerability as a gift brings to mind so many of the traits and states of being that mainstream white supremacy culture shuns. Why is it important to shift our framework from nuisance (or worse) to gift and opportunity? Why look for alternate paradigms and alternate ways of being? Climate scientists would say we don't have a choice. But there's a much more beautiful reason as well, linked to our evolutionary drive to be connected, a desire that has been driven out of mainstream, individualistic US cultural norms. As Rhonda Magee explains,

> through personal mindfulness practices, we can begin to ground, heal, and ultimately transform our personal sense of self, no longer clinging too tightly to a narrow and isolated sense of "I," "me," "my wounds," and the collective pain stories of "my people." We can begin to be able to infuse our experience of ourselves in culture, community, and context, with a sense of the valid, often painful experiences of others. And as we take in more of the whole, we grow. (Intro 34min)

Magee believes that healing happens in community, revealing to us "our common humanity and radical interconnectedness" (Intro 38min). Menakem has a similar idea that the paradigm shifts needed are not only individual work but also intergenerational, collective, and community based.

Given the revelations offered from the authors cited in this chapter, I ask community writing practitioners how we might create opportunities to inspire and engage community members in thoughtful discussion that will involve empathy, collaboration, and connection. Several of the authors agree that small, local groups are best for this kind of engaged work. Boulder-based writer Matthew King explains, "Anthropological experiments show us that, on average, any one individual can maintain stable relationships with 150 other people—a phenomenon known as 'Dunbar's number.' Beyond that social relationships begin to break down, undermining an individual's ability to trust and rely on the actions of others to achieve collective long-term goals." In light of this theory, in community writing projects, discussions and events with smaller social groups may have a better chance at encouraging people to engage and see themselves as having agency. "By re-shifting the conceptual frameworks about environmental concern as a *presence* or surplus of affect to be tapped, rather than an *absence* or lack to be refueled, new opportunities emerge for how communicators, educators, and activists can design and tailor messaging and outreach efforts" (Lertzman 146; original italics). In doing so, we strengthen the social and communal bonds that people crave.

Boulder-based climate scientist turned psychologist Jeffery Kiehl writes, "Our new story is one of global ecological emergence, providing us with a vision of the world as a single interconnected web in which we are co-creative contributors. Our challenge now is to work together to create this new story." The reality is that cultures across the world, including Native cultures in the United States, have lived for millennia though deep spiritual belief in a vision of the world as a single, interconnected web. So, it is not a new story that humans need but rather a (perhaps) new framing for people who live within white supremacist cultural ideologies and systems that eschew the collective for individual gain and power.

Relational accountability is an Indigenous worldview across cultures and guides how I understand *critical rhetorics of abundance*. Robin Wall Kimmerer explains, "The Indigenous philosophy of the gift economy, based in our responsibility to pass on those gifts, has no tolerance for creating artificial scarcity through hoarding. In fact, the 'monster' in Potawatomi culture is the Windigo, who suffers from the illness of taking too much and sharing too little" ("Serviceberry"). If we could only shift in mainstream United States culture to deconditioning ourselves from the belief that more equals better; that unlimited growth equates to progress; that happiness alone equals wholeness; that cheap, fast food can nourish us; and, for the academics reading this book, that our value as scholars is in where and how much we publish or in how much funding we bring in, or in all of the other ways we've been indoctrinated to conceive of abundance as simply meaning more for each of us as individuals. Indigenous scholars in and outside of Rhetoric and Writing Studies have been urging others across disciplines and communities to turn toward another way of being that values relationality, collaboration, and the knowledges and skills of others; care for others and ourselves in community; and giving as forms and actions of critical abundance.

The industrial food system could not exist as it is under these values. And if, as much as we are able, we do not want to participate in this system, we will need to seek alternatives for ourselves and our neighbors. A concept tied to that gratitude and abundance, Kimmerer suggests, is of mutual care. A beautiful anecdote demonstrates the difference between living with a scarcity mindset versus with an abundance mindset. She recounts,

> a report that linguist Daniel Everett wrote as he was learning from a hunter-gatherer community in the Brazilian rainforest. A hunter had brought home a sizable kill, far too much to be eaten by his family. The researcher asked how he would store the excess. Smoking and drying technologies were well known; storing was possible. The hunter was puzzled by the question—store the meat? Why would he do that? Instead, he sent out an invitation to a feast, and soon the neighboring families were gathered around his fire, until every last morsel was consumed. This seemed like maladaptive behavior to the anthropologist, who asked again: given the uncertainty of meat in the forest, why didn't he store the meat for himself, which is what the economic system of his home culture would predict.
>
> "Store my meat? I store my meat in the belly of my brother," replied the hunter. ("Serviceberry")

Think of how different the psychology is for this hunter than for, say, a survivalist hoarding for individual survival of an apocalypse or the people at a university who approve tenure guidelines that determine that collaborative writing is less valuable than single-authored publications or that storytelling or public writing are not valid forms of scholarship. The examples may seem far flung and different in scope, but they are all part of a culture of scarcity, individualism, and power *over*. Compassion, empathy, and humility do not equal weakness. Collaboration and community do not mean less for each individual. We can turn away from "What am I giving up?" and toward "What do we gain?"

Cognitive researchers, led by Daniel Kahneman, "have shown that we are consistently too loss averse. People care more about losing a dollar than gaining a dollar. About twice as much . . . But how has climate policy been framed to the public? As massively costly" (qtd. in Marshall 51), as loss and sacrifice: Give up your car. Give up beef. This "rhetoric of deficiency," as Hahn and Bruner label it, is not effective. Environmental research aligns with research on loss aversion in other disciplines. Marshall tells us, "Research suggests that people are more motivated to restore lost environmental quality than improve current environmental quality. There is therefore a potential to express climate change as an opportunity to **RESTORE PAST LOSS**, whether it is social (lost community, values, purpose) or environmental (lost ecosystems, species, or beauty)" (232, original bold and caps). Their question is whether climate policy can be framed as a gain instead of as purely a loss. Kimmerer asks, "Why have we permitted the dominance of economic systems that commoditize everything? That create scarcity instead of abundance, that promote accumulation rather than sharing? We've surrendered our values to an economic system that actively harms what we love. I'm wondering how we fix that. And I'm not alone" ("Serviceberry").

From Climate Change to Food Literacies

How does a community writing practitioner translate the findings—from climate communication; ecopsychology; critical race theories of trauma, grief, and wellness; and the other disciplines cited in this chapter—about how humans process difficult information in order to help communicate to multiple audiences in an ecology regarding any number of complex and troubling issues? I turn here to focus specifically on the topic of local food and how the above findings have helped me, my students, and my community partners to communicate about food systems.

The problems of the industrial food system are serious. Given what climate communicators and ecopsychologists teach, however, it will not be useful to hit people over the head with fear-based rhetorics or with "rhetorics of deficiency," which emphasize all we will have to give up eating now as well as what we may be forced to give up in the food-insecure future (Hahn and Bruner). The tension then, for Writing and Rhetoric scholars, teachers, and students, is to educate people about realities of the food system and to teach critical food literacies, while also making sure to offer them ways to become engaged and feel a critical abundance as a counter to deficiency and scarcity so that they don't shut down. Jenny Rice questions why it is so difficult to move people from shutting down to using that feeling in action: "In order to change people's minds, *you must also have an effect at an affective level*, which is more challenging" (97, emphasis added). Rice's insight is critical. Food literacy rhetoricians are battling evolutionary hardwiring, cultural conditioning, and powerful, deliberate marketing that push people to desire unhealthy, highly processed foods high in salt, sugar, and fat. Our arguments and interventions must be so affectively compelling as to outdo those powerful desires. There may be other powerful emotions related to food that we can appeal to instead.

According to Denver-based eco-hip-hop star, vegan chef, and food justice activist DJ Cavem, young people in particular are receptive to reconnection and more radical cultural change, often through art and music, which tap into affective responses in the brain. Cavem uses the power of music to spur young people, particularly Black and Latinx youth, into organizing and action through activist hip-hop. His latest EP, *BIOMIMICZ*, which comes with a seed packet, urges immediate action about food and environmental issues. On his website, Cavem explains, "*BIOMIMICZ* is about getting back to basics with a sense of urgency. . . . I hope my lyrics will inspire and educate. And I hope the seeds will be planted, literally." Through education and direct action, Cavem delivers a serious message to Black and Latinx communities through song and community connection, which will activate pleasure receptors in the brain.

Like the organics movement and plant-based eating movements, local food movements have been successful in marketing themselves as something people should crave. Tabloid magazines feature celebrities tending backyard chickens so that they can have pastured eggs; urban hipsters have rooftop gardens and blog about homesteading. But the movement must go deeper than a popular trend people may give up after a time, and this is where some local movements fall short of the essential antiracist and anticolonial connections being made visible though the leadership and vision of many Black, Indigenous, Latinx, immigrant, disabled, and other activists, farmers, and organizations.

In Black, Indigenous, Latinx, and immigrant communities that carry ancestral traumas and may have associations of working the land with slavery and genocide, for example, urban farms and places like Soul Fire Farm in New York, D-Town Farms in Detroit, and Frontline Farming in Denver are transforming the relationship with the land through stories of cultural reclamation and ancestral memory that encourage community members to get involved. Food localization, justice, and sovereignty movements are successfully shifting people's relationships with food through stories of resilience, and they are doing it through *critical rhetorics of abundance.*

When Leah Penniman, founding farmer of Soul Fire, tells the story of ancestral grandmothers "who braided seeds in their hair before being forced to board transatlantic slave ships, believing against the odds in a future on soil" (*Farming* 161) and therefore carrying their cultural knowledge with them, she makes an extraordinarily powerful case for seed saving, bringing back heritage seed varieties, and growing ancestral foods, even in the wake of profound trauma. When Chef Sean Sherman tells stories of his relationship with Native American foods in his award-winning *The Sioux Chef's Indigenous Kitchen*, he connects to his Oglala Lakota culture, explaining, "I tasted how food weaves people together, connects families through generations, is a life force of identity and social structure" ("Introduction"). Through story, he encourages Native Americans to look past the "fry bread and Indian tacos," which were not eaten by their early ancestors, to become deeply interested in ancestral foods, and in the process, to embrace "creativity as well as resilience and independence. Above all else, [ancestors] were healthy and self-reliant" ("Introduction").

How can we use the power of language to invite people into discussion, story sharing, and definition building around local food? What rhetorics can we use that might resonate with people on an affective level enough to elicit a change in behavior and consciousness? I suggest that we find an answer in *critical rhetorics of abundance*. Critical abundance is not about the piles of produce stocked in grocery stores, and it is not about the shelves packed with unhealthy products controlled by a handful of multinational corporations. It is not about individual accumulation. It is not about uncritical positivity. In other words, we need to argue that local food unites communities and nourishes us; it connects us to other people and animals and land, past and present.

Part of the positive response to local food movements, at least among middle- and higher-income populations, certainly comes from how it has been framed as a trendy thing to do. But there is something much more profound happening too, and it connects the Oakland homesteader keeping bees in their backyard to the mother growing food for her family through a Detroit community garden project to the child enjoying vegetables for the first time because he grew them in his elementary school garden in Boulder. During our interview for this book in 2020, I asked then–Boulder County Farmers Markets executive director Brian Coppom, to elaborate on his idea that local foods remind people of our connection to food, one another, and the earth. His response was simple but profound: "One of the things that we try to do is invite people to experience it. I'm certain that people are more swayed by the experience of something than by the intellectual understanding of something. So that's why we love them to come to market just to see if there's a little something, one of those brain cells that's kinda like, 'Oh, like I remember this connection.' And you know, maybe that stimulates a little cellular conversation in the brain, and it starts growing. But I think shaming is a bad way to do it. I feel like maybe one of the pieces of the farmers market that works is the farmers tell their stories directly to our patrons. So, we're experiencing it directly through stories, through being there, hear, smell, feel."

Chef Sean Sherman takes it even further to suggest a critical decolonial element: "[The Indigenous diet] is a diet that connects us all to nature and to each other in the most direct and profound ways" ("Introduction"). What Coppom has called a "primal connection," others may think of as ancestral and decolonial in connecting foodways, which were destroyed or forcibly left behind as part of the colonial mission, to a past, present, and future that is not just individually resonant but culturally critical for survival.

If local food movements can tap into that deep spiritual, ancestral yearning that so many humans feel for connection to the earth and other beings, for community, for transparent, ethical food choices that keep us, our families, our cultures, and our communities healthy, they might succeed. This will be because these local movements, directly tied to place and relationships, offer a profound affective critical abundance that big ag and industrial food companies can never achieve. It is also why an *ecological community writing* methodology includes histories of land, place, and self, as well as ancestral stories and legacies. These elements, as much as farms, restaurants, markets, nonprofits, and other entities, make up a local food ecology. And when we ask people through our projects and events to write their definition of local food, we can encourage them to consider all of these elements in their definition building.

Rhetorics of Deficiency and Abundance in Boulder County

During the same period of time that I was teaching food-focused writing courses at the University of Colorado Boulder, about two dozen local leaders—including Boulder's mayor, a county commissioner, officials from several city and county government offices, executive directors of several nonprofits including the Boulder County Farmers Markets, CEOs of food-related businesses, representatives from the Boulder Valley School District and University of Colorado Boulder, and farmers—agreed that local food should be a priority for Boulder County. Spearheaded by the bold vision of then–Boulder City Councilman Tim Plass, this group of leaders formed a working group, Making Local Food Work. Its mission, until the group disbanded in 2019 because of the demands of each person's work requirements with their affiliated organizations, was "to balance our food system by promoting the increased production, consumption, and preservation of regional and local food." The board originally used a shared definition for local food based solely on miles—"local food" is anything grown or produced within the Boulder County foodshed. Like a watershed, a foodshed refers to a physical region through which food moves from place of production to place of consumption.

At a 2016 Local Food Summit in Boulder, one of my community partners introduced me to Plass. We talked with one another about our respective projects and shared interests, and he invited me to attend a name-changing celebration later that week for Making Local Food Work's transition to The Shed: Boulder County Foodshed. He also invited me to present my classes' findings from the past several semesters of projects at the next Shed board meeting. He thought the research might be useful to the group as they discussed future goals. This was the connection my students had hoped for when they asked me to find a way to share their research on definitional confusion about what local food is and means.

Once I presented to The Shed board, we discussed what next steps might be given the definitional complexities my students and I had uncovered. I suggested the possibility for a collaborative community literacy campaign about local food co-created by CU students, The Shed board, and the various organizations and businesses the board represented. The collaboratively determined food literacy goals were that more people understand what they are eating, what systems their purchases support, and what the challenges are for scaling up local food production and consumption in Boulder County.

The Shed board members invited me to serve on the board and, eventually, to chair the Education and Promotion Committee, work that allowed me to witness the complex rhetorical situation firsthand. In an email that Tim Plass sent to me after reading a draft of this chapter, he explained the reasoning behind the organization's name change as aligning with desire for consideration of a broadened definition for "local food":

> The foodshed concept was meant to challenge people to think about where our food comes from and get away from the limiting parameters of "local" that you discuss at length in your book. So, perhaps a West Slope peach could be local, but not something like West Slope kale, for which much closer, better options exist. If we could succeed in making Boulderites have an awareness of the food they choose, be informed, and make intentional choices that support community (or their own) values that would be a huge step forward! So, the whole idea of a foodshed is a more nuanced, encompassing way of looking at the issue: challenges and opportunities. (Plass)

The board members sometimes had polarized positions on issues related to agricultural and environmental policy. Even within The Shed's small number of board members, the dilemma of ethics/environmental impacts versus food miles was prominent and multifaceted. As we discussed how to be most useful as a group, we acknowledged how extraordinarily complicated it is to support local food production and consumption. We hoped to dig into the issues to best know how to intervene. We talked about some of the findings I mention earlier in this chapter from trauma, communications, and ecopsychology experts, and we ultimately drew from them to help develop the frames we could use in educating the public.

Joining The Shed board and chairing the Education and Promotion Committee let me experience the rhetorical interplay among the board members, all of whom cared deeply about local food production, as each defined it. Some, who worked more closely with the conventional farmers who lease Boulder County Open Space land, would at times draw on "rhetorics of deficiency" (Hahn and Bruner) in their arguments for why conventional agriculture and GMOs are an important part of a sustainable food system along the Colorado Front Range: Farm families, many of whom have farmed the land for several generations, would go bankrupt if they were forced to switch to organic production; there is not enough water in Colorado to support all farmers growing organically; many conventional farmers have private contracts with companies requiring certain types of seeds; there is high turnover among organic farm workers because the job is so labor intensive; the market demand isn't there for small-acreage crops such as heritage wheat, pinto, and garbanzo beans (Booth). Their framing of the issues raised serious considerations.

In fact, the majority of articles written a few years later about the Boulder County commissioners' December 2021 vote to reverse the previous decision to phase out GM crops on leased Open Space agricultural lands used similar rhetorics of deficiency and scarcity. These are common frames in local food rhetoric here in the county and elsewhere. As a newspaper article about the reversal tells the story of a GM commodity crop farmer, Paul Schlagel, readers find rhetoric such as "But he's not willing to lose his family farm raising crops that people won't pay for," and "the farmers convinced Boulder County leaders that they couldn't survive while adhering to prohibition of genetically engineered corn and pesticide coatings using controversial neonicotinoids" (qtd. in Booth). Schlagel argues, "It's hard for people living in the city of Boulder, and even on the western side of Boulder County, to understand what real agriculture even looks like," and, he maintains that the crop bans create "a mandate from somebody that we're going to transition to something better that is very naive. Because if there was something better, we would already be there" (qtd. in Booth). As a county commissioner at the time argued, "People have this romanticized view of what farming should be. We have thousands of acres of irrigated cropland in Boulder County that are going to be farmed with row crops ... and there is absolutely no way that all that land is going to become little 10, 15, 20-acre organic plots" (qtd. in Booth).

In my understanding of where some county officials, farmers, and others derive their arguments, I draw on Paul Feigenbaum's theory of adaptive rhetorics and "the adaptive function," which can make people reticent to accept counternormative ways of looking at and being in the world (*Collaborative Imagination* 18). There is an adaptive function of corporate capitalism that has lulled people into believing in the economic myth of infinite growth and prosperity, and that abundance equates to quantity rather than quality. Adaptive rhetorics encourage complacency or apathy or the belief "that's just the way it is." Anything else, as climate scientist turned psychoanalyst Jeffery Kiehl suggests, is virtually unfathomable: "We literally cannot imagine" anything else. That is the function of adaptive rhetorics that, Feigenbaum argues, keep people from striving for what Miles Horton called the *"ought to be"* (32). Multinational food and seed corporations invoke adaptive rhetorics through ethics of care: feeding the world; ethics of innovation: making food as cheaply and quickly as possible; ethics of technological supremacy: farm as factory and animals as product. In succumbing to the adaptive function of corporate capitalism, people literally buy into it.

On the other side of the spectrum, many county residents, especially those who live in the city of Boulder, used fear-based rhetorics to argue against allowing GM farming. As many county officials argued, the public sometimes had fears that were not always backed by scientific study about the safety of consumption of GM crops and about the environmental standards for farming GM crops.

Community writing practitioners can illuminate the rhetorical patterns at play that function to keep people from interrogating these discourses, and we can instead encourage community members to actively engage. For Rice, that literacy manifests through a deep inquiry process, where people learn to recognize and question the adaptive rhetorics that capture the public imagination. My fellow community members and I might seek answers to important questions about, for example, why one in eight Boulder County residents, 41,000 people, is food insecure; why farmers exist on such razor-thin margins between economic viability and bankruptcy; why there are so few BIPOC farmers in the county; why Native foodways are not discussed more often in considerations of what to grow as food; what independent, peer-reviewed scientific studies indicate about safety of GMOs. These are critical questions to pursue. I push beyond Rice's focus on inquiry as the end goal, however, because if the findings I present in this chapter teach us anything, it is that asking these questions without providing strategies for meaningful collaborative engagement may not ignite change and may, in fact, cause further overwhelm, anxiety, or disengagement. The Shed needed to create a way of communicating and framing information, one not focused entirely on logos-driven inquiry and not focused entirely on fear-based pathos.

Critically framed *rhetorics of abundance* can harness the positive emotions people associate with local food, such as those my students heard from farmers market patrons, while also encouraging more substantive inquiry, reflection, and action into a complexity of emotions on both the personal and community levels. Climate communicators and ecopsychologists also teach that collective action is more emotionally powerful than individual action, so The Shed, my students, and I developed ideas for bringing people together in community. As I will discuss in the next chapters, thinking in terms of *ecological community writing* through distributed writing opportunities helped me work with other board members to reach as many community members as possible through public events and projects that encouraged engagement. Thinking in terms of *ecological community writing* also helped me organize an interdisciplinary coalition of faculty, graduate students, and undergraduates from across campus as an ideal way to help build the local food ecology.

To break from adaptive rhetorics that lull us into complacency and to instead move us to action, we need "radical imagination," a concept, rooted in Black feminism, a practice for "imagining a world that doesn't exist" in order to construct more just worlds in the present (Aloziem). The concept, which has been used by visionaries such as Angela Davis (e.g., *Abolition* and *Politics*), Jacqueline Jones Royster ("Disciplinary"), adrienne maree brown (e.g., *Emergent* and "Radical"), and Carmen Kynard (e.g., "All I Need" and "Oh No"), is a collective practice and a process. In *Freedom Dreams: Black Radical Imagination*, Robin D. G. Kelley asks: "How do we produce a vision that enables us to see beyond our immediate ordeals? How do we transcend bitterness and cynicism and embrace love, hope, and an all-encompassing dream of freedom?" Our community-based work of envisioning and co-creating a more just food system can be collectively guided by and called into being through joyful practice that moves our communities into active imagining of alternate possibilities to what currently exists. And as all of the scholars named in this paragraph attest, this imagining does not imply a turning away from violent and traumatic historical and current realities. Radical imagination faces injustice and crisis head on.

This imagining can address Schlagel's argument that "if there was something better, we would already be there" and the county commissioner's position that "there is absolutely no way that all that land is going to become little 10, 15, 20-acre organic plots" (qtd. in Booth). When people argue that there cannot possibly be another way to farm Open Space land, because what we've tried so far to convert to organic or regenerative production hasn't worked, it does not mean there is no other way. We just haven't dreamed it yet. We're so used to what is currently, that it seems inevitable.

It's a similar issue in academia. When we value certain knowledges, when we create hierarchies of power, we are living in the imagination of the (almost exclusively) white men who established the rules. And so many academics of all genders and races maintain the existing structures because they seem inevitable. Radical imagination asks that we collectively begin with "What if?," the question Royster asks herself in her work to contest established patriarchal and racist norms in Rhetoric and Composition (Royster, "Disciplinary" 150), and move into the "how?"

It may be that the little organic plots are somehow possible on large parts of the farmland in Boulder County; or perhaps they are not, and something even more extraordinary can be imagined into being. What are the many ideas "What if?" conversations could generate? As adrienne maree brown reminds us, "We need to move from competitive ideation, trying to push our individual ideas, to collective ideation, collaborative ideation. It isn't about having the number one best idea but having ideas that come from and work for more people" ("intersecting"). In Boulder County and across the country, there are innovative farmers, activists, organizers, and government officials asking the "what if" and "how" questions every day. As others learn from them and also ask, "what if?" and "how?," collective and collaborative ideation happens. This is what *distributed definition building* encourages.

"Empathy calls for imagination … Imagination stems from a connection with my own pain," Breeshia Wade explains. In this vein, it is important to recognize with empathy the fear, pain, and anger that conventional farmers may feel as they contemplate giving up their way of farming. We can also recognize with empathy the frustration, fear, and anger community members may feel for the resistance to change, especially if they feel that their own and their family's health and the health of wildlife and the land are threatened. And we can acknowledge our locale's limits in terms of current production, length of the growing season, water availability, soil quality, and current rhetorical frames being used to guide the conversations, while also actively imagining more just futures for the county's farmland, farmers, animals, and community members. "How do we survive these falling systems?," brown asks. "Especially when many of them need to fall? How do we prepare for the opportunities in collapse?" ("Darwin"). Framing an issue like local food through the lens of *critical rhetorics of abundance*, we may believe industrial food production in its current forms and practices is ultimately a system that will fall. But we have to be willing to break from the indoctrination that certain imaginations (often framed through a white, cis, hetero, ableist lens) are more important and valid. Then, as brown urges, we radically imagine opportunities.

Community-engaged practitioners in Rhetoric and Writing Studies can foster ways for distributed imagining through *ecological community writing* projects and events rooted in *critical rhetorics of abundance* that turn away from adaptive and deficiency rhetorics. In the next chapter, I explore ecological writing theories and systems theory, which tap into this notion of the power of the collective. Then, in chapters 5 and 6, I offer concrete examples of how my students, community project partners, and I engage fellow community members, using several of the theories presented in this chapter as applied to community food literacies at the factual, rhetorical, and affective levels to encourage engagement.

4

Writing Abundant Ecologies

FROM THEORY TO COLLABORATIVE LOCAL RESEARCH

When you step on one strand of a spider web, it all moves.
—PATRICK SPEARS, QUOTED IN WINONA LADUKE, *ALL OUR RELATIONS*

We are a part of everything that is beneath us, above us, and around us. Our past is our present, our present is our future, and our future is seven generations past and present.
—HAUDENOSAUNEE TEACHING, QUOTED IN
WINONA LADUKE, *ALL OUR RELATIONS*

The trauma researchers, climate communicators, and ecopsychologists I discussed in the previous chapter, as well as Black feminist theorists such as Angela Davis, Jacqueline Jones Royster, adrienne maree brown, and Carmen Kynard, have emphasized the power of radical imagination in moving people toward transformational change and the embrace of complex emotions of love, grief, anger, pleasure, and hope. People need to feel that they can contribute to the extraordinarily difficult work that is called for through collaboration and creativity, tied to our evolutionary drive for group connection. This scholarship influenced the development of *critical rhetorics of abundance* for community food literacy projects. Even with abundant and collective framing for an individual project or event, however, there is still the question of how to spread a message beyond that single instance, of how to position that isolated work to move and reverberate through an ecology.

When we assess our students' community-engaged writing or create writing projects in our communities, we may not account for the very complicated ecologies in which our students' and fellow community members' texts function, circulate, and can be transformed beyond the original authorial intent. Although since the social turn in Composition Studies, we as a field have embraced the importance that people write *in context*—a concept that has helped community writing practitioners to explain the benefit of engaged writing courses—we often lose track of the students' or community's writing once it is produced. In other words, in a community writing course, for example, a final project may be a written product produced for a partner; however, once a grade is assigned or a project ends, we often do not continue to study the writing's contextual nature. Rarely do we see that writing again. Even more rarely do we and future students study its trajectory from classroom through communities. But what happens to it? How does it interact and intersect with, diverge from, morph into other writing and ideas as it or its message circulates? In other words, how does community writing become part of and impact the local ecology?

Indigenous theories of relationality, ecological writing theories, and circulation theories all offer partial answers to these questions. In this chapter, to understand possibilities for encouraging what I consider a saturation of local food rhetorics across multiple audiences and spaces, I delve into some of the theories and scholars who have guided my thinking about the capacious possibilities for writing and rhetoric to guide work in *ecological community writing*. In the second half of the chapter and in the chapters to come, I turn to praxis, putting theory into action.

In Rhetoric and Writing Studies, ecological theories of writing and new materialist rhetorics have become popular and gained traction, and scholars who affiliate with ecological writing studies and new materialism often trace connections to Bruno Latour, Gilles Deleuze and Félix Guattari, and other white, male, European thinkers (e.g., Gries, *Still Life* 4–5). Only recently have some non-Native scholars in this area of the field acknowledged the centuries of Indigenous knowledges that teach of kinship and "all our relations" between beings (Kimmerer, *Braiding*; LaDuke, *All Our Relations*; M. Powell). Beyond human-to-human connection, when we consider connections with other-than-humans through Indigenous, ecological, and new materialist theories, we can begin to understand the ways in which we act, are acted upon, and act with other people, places, and beings. We live within vast and always-changing material, rhetorical, linguistic, and affective ecologies. These ecologies are, by their nature, relational (M. Powell; Kimmerer, *Braiding*).

The omissions of Indigenous thought happen in posthuman and new materialism studies in multiple disciplines. Métis anthropologist and scholar Zoe Todd points out that Indigenous scholars have long written about the animacy of other species and objects before higher education's fascination with the ontological turn. Sisseton-Wahpeton Oyate scholar of Native studies Kim TallBear said in a 2011 address entitled "Why Interspecies Thinking Needs Indigenous Standpoints," "The academy is now being infiltrated by *non-indigenous* voices articulating the idea that life/not life is too binary and restrictive. This indicates greater scope at this moment in history for bringing indigenous voices to the conversational table" (original italics). The scarcity of acknowledgment or inclusion of Indigenous voices in many of these academic conversations has led to critiques for lack of more expansive framing, citation, and epistemological lineage. White scholars Jerry Lee Rosiek et al. explain why this is so important for scholars of Euro-American heritage to address—not only to bring into dialogue two strands of thought that share similarities but also because failure "to acknowledge and seriously engage the Indigenous scholars already working with parallel concepts, end[s] up reinforcing ongoing practices of erasure of Indigenous cultures and thought (Ahmed, 2017; Deloria, 1999b; Todd, 2016; Tuck, 2014; Weheliye, 2014)" (Rosiek et al. 332). It is up to "Eurocentric scholars," Rosiek et al. continue, "to recognize the responsibility to engage Indigenous thought and traditions and to do so in light of the history of colonization, displacement, and genocide" (Rosiek et al. 335).

In Rhetoric and Writing Studies, Malea Powell and Andrea Riley Mukavetz, among others, have raised concerns with object-oriented ontology and ecological and new materialist theories of writing that ignore non-European rhetorical traditions, including Indigenous concepts of relationality, constellations, and webs of connection (as referenced by Spears and LaDuke in the epigraphs of this chapter). In this chapter, while I include theories of ecocomposition, new materialism, ecological rhetorics, circulation studies, and the associated scholars as parts of the discipline who have helped to shape my thinking and my projects, I also acknowledge limitations and omissions in the scholarship. As I discuss in detail in chapter 6, I am guided by Indigenous and Native scholars who call for a more capacious reading of and theorizing of ecologies and lineages of thought for the field at large.[1] I am sure that I too will leave out critical voices because of my own limited knowledge, and I hope others will continue to right these disciplinary and epistemological omissions.

Bringing multiple strands of scholarship and ways of being together when considering *distributed definition building* in an *ecological community writing* methodology is essential in questions of accountability and repair because of the on-the-ground consequences of omission. Epistemic justice helps community writing practitioners ask questions of who is and is not present, who is and is not yet considered part of an existing ecology, whose knowledges and definitions are useful and whose are ignored or need to be addressed through further literacy projects.

A common community writing model might consist of one class that works on one or more projects with one partner but does not take into account the whole ecology in which we and our students write. In an ecological or new materialist writing model, we might think about how actors, actants, and writing interact and circulate in an ecology, but how do we encourage justice-focused ecological work in practice? In Indigenous writing models, we might consider stories and relationships between beings not just as conceptual ideas but as ethical ideas—as embodied spiritual and cultural ideas that ecological writing studies does not often get at. How do non-Indigenous scholars encourage story and relationality in community writing spaces while mindful of cultural appropriation and anticolonial justice work? As I discuss in the following chapters, we can find one possible answer in *ecological community writing*, which offers a confluence of theory and practice to create abundant

---

1 Indigenous scholars are not alone in pointing to lineages of thought ignored in traditional rhetoric and writing programs that look to classical lineages as the foundation of the field. Significant works in Rhetoric and Writing Studies by scholars of color also center rhetorics and methodologies outside of the Greco-Roman and white European tradition to expose the "disciplinary landscaping" (Royster) that has been sparse on acknowledging and citing this important work: from Richardson and Jackson's edited collection *African American Rhetoric(s)*, Royster's *Traces of a Stream* and Kynard's *Vernacular Insurrections: Race, Black Protest, and the New Century in Composition-Literacies Studies*; to Ruiz's *Reclaiming Composition for Chicano/as and Other Ethnic Minorities* and García and Baca's *Rhetorics Elsewhere and Otherwise*; to Kinlock's *Harlem on Our Minds*, Pritchard's *Fashioning Lives*, and Hsu's *Constellating Home: Trans and Queer Asian American Rhetorics*; and to work to expand methodologies and pedagogies in A. Martinez's *Counterstory*, Roossien and Riley Mukavetz's *You Better Go See Geri*, Lockett et al.'s *Race, Rhetoric, and Research Methods*, King et al.'s *Survivance, Sovereignty, and Story: Teaching American Indian Rhetorics*, and so many more important scholars and works. While the field in more broad terms has finally and recently undergone a shift to celebrate and cite these texts, it is slow coming, and ecological Writing and Rhetoric Studies mirrors that slowness.

rhetorics and to recognize ourselves as teacher-scholar-organizer-activist, or any of the myriad ways we define ourselves, whose work is not only to function in an ecology but to help create that ecology.

As I began to work with my long-term partner, The Shed: Boulder County Foodshed, we realized that there were duplications of work happening in organizations that did not know of one another; there were gaps in work because people did not have a sense of the scope of food-related work happening in the county; and even among organizations that knew of one another, there was not always a sharing of resources, data, and knowledge. Part of our mission became to teach people about local food, recognize who is already working on local food issues with whom we can connect, move ideas and communication about local food throughout the county, connect parts of the ecology not currently connected, and all the while, encourage people to care enough about local food to act in support of producers. This work demands an ecological, distributed approach.

Indigenous concepts of relationality and other-than-human kinship, ecocomposition, ecological writing studies, new materialism, systems theory, rhetorical contagion and virality, and circulation theories have deeply influenced my understanding of how we can conceive of writing and rhetoric's force and vitality in the world, of the fluid and alive nature of rhetoric and writing. Because the theories are sometimes abstract, community writing practitioners need to figure out how they can be useful and play out in community literacy and community writing projects. After first describing several of the key concepts of the theories named earlier, I turn in the second half of the chapter to several examples of how I put into action these theoretical concepts through an interdisciplinary research project and several collaborative endeavors with The Shed and other project partners. I hope that these projects, as examples of my *ecological community writing* methodology, offer a praxis of the on-the-ground work we can do in our communities to shape how definitions, and therefore community literacies, are generated.

Ecological and New Materialist Rhetorics

Ecological writing studies has, for more than three decades, posited writing as part of an ecology. In the European-American, predominantly white tradition of scholarship that gained popularity in the 1980s and 1990s, Marylin

Cooper, Margaret Syverson, and others theorized, for example, how writing students and classrooms are ecologies, and they applied theories of ecological systems to students and their composing processes. Early ecocomposition wrestled with how these concepts should shape the work faculty (and sometimes students) do. Ecocomposition gained popularity in the early 2000s, though Sidney Dobrin and Christian Weisser trace it back to the 1970s. Their 2002 book *Natural Discourse* offers a preliminary definition from Dobrin and Weisser: "[E]cocomposition is the study of the relationships between environments (and by that we mean natural, constructed, and even imagined places) and discourse (speaking, writing, thinking)" (6). One strand of ecocomposition has tended toward courses that focus on eco-pedagogy—teaching writing about environmental issues. Dobrin and Weisser's early writing argues for "activist intellectuals" (*Natural Discourse* 87). They explain that "ecocomposition must include a component of activism and participation that moves beyond the classroom space" (Weisser and Dobrin 7). Because Rhetoric and Writing scholars are "uniquely positioned to offer our abilities as rhetors and writers to local organizations and to be vocal in local debates," "ecocomposition must be an active *praxis*" (*Natural Discourse* 102, 115). Along these lines, both scholars and students would recognize themselves as subjects among many within larger "systems, locations, and environments" (83) and that "local activity directed at more confined publics may begin to affect larger publics as the effect of local activism disperses through a range of discursive environments" (97).

As his theories evolved, Dobrin has raised the question, "What is writing?" (*Postcomposition* 10). The strand of ecocomposition that Dobrin later developed, which he has called "postcomposition," is really a questioning of the very nature of what writing is. Dobrin ultimately distanced himself from the eco-pedagogy connection to ecocomposition and wrote more explicitly on writing *as* system. "Ecocomposition's ecology," he explains, "has been flat, limited to an ecology bound by metaphors of a web, some consideration of the 'placeness' of writing, and an overexerted attention to nature writing and environmental politics instead of any serious attention to the ecologies of writing as system" (141). Dobrin posits instead "an ecological/networked theory of writing that does not rely upon subjects as a principal tenet of the theory" (4).

In terms of community-based work, this of course doesn't mean that there aren't individual writers anymore but rather that we're all always connected to other writers, organizations, and things that impact and shape what we produce and that may use what we produce in unforeseen ways. Tom Deans agrees with "the need to understand (and help students understand) writing as deeply embedded in culture and context; the need to conceive of writing as a social, relational, and multimodal process rather than as just an individual and textual one" ("Sustainability" 103). Marylin Cooper, who influenced Weisser and Dobrin, has urged us not to use the concept of ecology as a metaphor. She writes that "the systems that constitute writing and writers are not just *like* ecological systems but are precisely ecological systems, and that there are no boundaries between writing and the other interlocked, cycling systems of our world" ("Foreword" xiv). Jenny Edbauer's study of rhetorical ecologies in her article on the spread of the Keep Austin Weird phenomenon and Laurie Gries's *Still Life with Rhetoric*, which tracks the rhetorical evolutions of the Obama Hope image throughout the world, are two of the many examples of disciplinary study of rhetorical ecologies. Gries argues that "[r]hetorical discourse emerges from a dynamic assemblage of agents involved in ongoing intra-actional performances. Within this assemblage, people are just one agent among a host of other human *and* nonhuman entities that have potential to catalyze change" ("Agential" 69).

Community writing practitioners might ask, given these theories in Rhetoric and Writing Studies, what the material bodies are through which both writing and rhetoric manifest, remix, and circulate in our communities. The question of what "local food" means in Boulder County is very much rooted in the Boulder County ecology, and the points of dissonance and agreement in its definition are connected physically, culturally, historically, affectively, and rhetorically to Boulder County. In any study of the local, we can study the complex relationships between the parts as they relate to the whole. I see writing as the connector between parts, and our work can help to generate those connections.

Terroir

A concept that has helped concretize these ideas for the work of new materialists and rhetorical ecologists is *terroir* from the French wine-tasting concept that each wine has a distinct taste and aroma developed in connection to where and how the grapes are grown. The concept has been adopted for food studies as well and has been picked up in rhetorical studies. The placeless, fast, conventional, replicable food that comprises the majority of most Americans' modern diet has no identifiable *terroir*—the particular signature of a place as expressed through its produce and products. The soil, water, farming practices, climate—all impact taste and embody *terroir*.

The concept of *terroir* has evolved to refer to the "culture and history of a place": "The importance of the *terroir* concept is its ability to link directly the growing practices and the physical environment, the cultural associations of the food, and the place where it's grown" (Gottlieb and Joshi 191). As an example, in her analysis of Texas BBQ and *terroir*, Anna "Amy" Young explains, "the notion that land, shaped by culture, by natural environment, by history, by politics, and by those who tend to it, gives rise to a specific, identifiable, and distinct flavor and style"; "in a rhetorical sense, terroir *means* and *does*" (40, 45).

Because *terroir* has become a popular trope in ecological writing and food rhetorics, I want to consider how Rhetoric and Writing scholars can learn from such concepts. *Terroir* is a constellation of place, culture, taste, soil. I'm most interested in its connection to locality and to local food systems in terms of connections to or rejections of justice-focused work. Every manifestation of local food will be different depending on its locale.

Linking back to the "On Lineage" section of this book and the idea of a placeless curriculum that can be transferred anywhere, *terroir*, in relation to faculty and students, connotes the opposite—courses and research that are of the place itself, that develop in relation to land, culture, community conversations and knowledges, soil, history, and on and on. All of these factors can shape a curricular *terroir* and community writing research responses.

In a similar way, the concept of local food is both concrete and abstract, embodied in rhetorical choices and embedded in language that circulates locally and can impact policy and community action. It is tied culturally to the history of the land and who lives on it; to the historical work to protect Open Space; to the plants, seeds, water, and soil for what they can teach us about what kinds of food systems are possible. Jane Bennett asks, "How would political responses to public problems change if we were to take seriously the vitality of (nonhuman) bodies? By 'vitality' I mean the capacity of things—edibles, commodities, storms, metals—not only to impede or block the will and designs of humans but also to act as quasi agents or forces with trajectories, propensities, or tendencies of their own" (viii). The capacious definition of *terroir* as not just a concept but as a rhetorical force relates to the ways in which I think of the multiplicity of definitions for "local food."

I consider the opportunities for linguistic choices and definitional decisions as locally inflected, asking how Boulder County (its people, its physical environment, its laws, its history, its affect, and more) "writes" meaning. I will offer in chapter 6 an idea, rooted in Indigenous knowledges, that we shouldn't separate our definitions of "local food" from what the land itself "writes" as possible. In what ways do the land, the soil, the plants and animals "write" our answers to what kinds of local food production are possible in the county? Or in what ways do the histories of Open Space preservation and community activism as well as settler colonialism and destruction of Native food cultures "write" the current affective and rhetorical landscapes for what actions community members will accept in support of local food production? With these questions in mind, we might consider the functions of writing not just in print or image but as threads in an ecological system.

Writing in Systems

The rhetorical concepts of relations, actants, assemblages, ecologies, *terroir*, and other-than-human authorial agency all relate to systems thinking, an understanding of which can help shape community writing projects in order to create impact. In *Thinking in Systems*, Donella Meadows explains, "A system is a set of things—people, cells, molecules, or whatever—interconnected in such a way that they produce their own pattern of behavior over time," and the reason it is hard to apply what works in one system to another is that "the system, to a large extent, causes its own behavior! An outside event may unleash that behavior, but the same outside event applied to a different system is likely to produce a different result" (2). A system is like a *terroir*, grounded in locality, and "an interconnected set of elements that is coherently organized in a way that achieves something ... a system must consist of three types of things: *elements, interconnections*, and a *function* or *purpose*" (Meadows 11, original italics). The concept of the relationality here is useful—that things act together in ways they would not separately. For this reason, it is important that community writing practitioners work more in terms of systems approaches than individual manifestations of issues or problems.

The theory of unintended consequences teaches that because big, "wicked" problems are "embedded in larger systems, . . . some of our 'solutions' have created further problems," as is apparent in the industrial food system that I discussed in chapter 1 (Meadows 4). In one of my favorite scenes of the documentary *Food, Inc.*, self-proclaimed "beyond organic" farmer Joel Salatin says,

> I've always been struck by how successful we have been at hitting the bullseye of the wrong target. We've learned how to plant, fertilize, and harvest corn using global positioning satellite technology, but nobody sits back and asks, "should be we be feeding cows corn?" We've become a culture of technicians. We're all into the "how" of it but nobody is stepping back and saying, "but why?"

As Salatin's insight suggests, the shortsightedness in so many decisions related to industrial food production shows the dire consequences of not thinking systemically. A goal in *ecological community writing* is to encourage people to step back and ask, "but why?"

Winona LaDuke offers an example related to cattle that can be understood in terms of problems at a systems level that have present-day interconnections as well as historical ones. The buffalo killing of the 1800s, accompanied by white settler colonial seizure of land for cattle production as a way "to settle the vexed Indian question" by starving them off their land, has led to current consequences of the massive factory farming of cattle in the West (*All Our Relations* 143). These historical realities have led to present-day unintended consequences: "Industrial agriculture determines the entire ecosystem, from feed crop monoculture to feedlots, from unpriced public grazing permits (a holdover from old reservation leases) to drawdown of the aquifers, agricultural runoff, and soil erosion. Much of this is for cows" (145–46). The interconnections between settler colonialism, industrial meat production, environmental degradation, corporate capitalism, and cultural genocide demonstrate the importance of thinking in systems rather than in terms of individual, present-time manifestations of an issue.

Stepping back to see the "why?" means taking a historical view that sees how multiple and continuous decisions have led to the current food landscape. Systemic problems need systems-level answers. "Systems happen all at once," Meadows explains, and "are connected not just in one direction, but in many directions simultaneously" (5). This is why an isolated approach to one problem often causes other problems. In *Systems Thinking for Social Change*, systems theory expert David Peter Stroh agrees that it is critical to use systems thinking in trying to address complex problems. "Conventional or linear thinking works for simple problems," he explains, but not for complex social problems (14).

In Boulder County, for example, farmers have been vocal about how water availability, soil health, and land-use codes that impact the number of allowed events or kinds of structures allowed on a farm are multiple layers of limits that impact the local food system. Meadows posits that "Intangibles are also elements of the system" (13), and here we can connect what we learn from ecopsychology regarding affect and consider it from a systems—and ecological—perspective. The *affect* of local food is part of the ecology. Pride, joy, fear, nostalgia, anger, environmental concern, grief are all parts of the ecology. In a quest to find and generate the interconnections, community writing practitioners might ask whether writing itself could be one of these interconnectors; and if writing, capaciously defined, *is* a connector in an ecology, we can use it to help alter the ecology.

Meadows suggests that we "look for leverage points—places in the system where a small change could lead to a large shift in behavior" (145). "Starting with the behavior of the system," she explains, "directs one's thoughts to dynamic, not static, analysis—not only to 'What's wrong?' but also to 'How did we get there?' 'What other behavior models are possible?' 'If we don't change direction, where are we going to end up?' And looking to the strengths of the system, one can ask, 'What's working well here?'" (171). These questions can bring people from surface-level thinking about individual manifestations of an issue, about why they like local food, for example, to much deeper questions that help them reflect on the interconnectedness of root causes of issues that make local food production and accessibility so vital. If community literacy practitioners were to build projects deliberately thinking about systems, we could consider how a systems approach would shape our pedagogy, research, and outreach as we worked toward what V. Jo Hsu calls "the story that will bring us together" (194).

We can consider how writing and rhetorical intervention could prove useful in systems-level work, even when problems seem so vast and out of our control. Many community literacy scholars are doing work that I would consider ecological and focused on systems, though they may not use that specific vocabulary. Two examples of many possibilities follow to show the constelled or ecological work community literacy scholars have been doing. In Elaine Richardson's significant work with Black women's and girls' literacies, for example, she studies the "constellation of African American cultural identities, social locations, and social practices that influence how members of a discourse group make meaning and assert themselves socio-politically in subordinate as well as official contexts" as she explores various spaces that try to define Black women and girls: "the streets, the workplace, the school, or the airwaves" ("My *Ill* Literacy" 755). Richardson is studying an ecology of writing, language, and literacy. Another example of taking an ecological or constelled approach to projects is Joyce Rain Anderson's multiple courses and projects that she has fostered in collaboration with local, Native populations—from establishing a powwow with the Massachusetts Center for Native American Awareness to integrating Indigenous programming and pedagogy into her university's strategic plan to starting a three sisters (Indigenous name for corn, beans, and squash) garden on campus to creating cornhusk dolls with her students while telling Corn Mother stories (164–67). Ecological visioning of not only issues and populations but asset-based project development makes good sense given the complexity of systems.

As I discussed in the previous chapter, climate communicators agree, though perhaps not always for ancestral, cultural, and decolonial reasons, that the power of story cannot be underestimated in helping people understand the complexity of feelings (as opposed to easy, naive, uncritical feelings) associated with loss and in bringing people together for collective action. We can take our cue from our many colleagues, particularly in cultural rhetorics, Indigenous rhetorics, and antiracist rhetorics, who offer alternative methods and methodologies that encourage story and counterstory, connections, multiplicity, and experimentation (e.g., A. Martinez; Kynard [e.g., *Vernacular*]; Hass; Hsu; Ruiz; Driskill; E. Richardson [e.g., "She Ugly"]; Ore et al.). I offer these examples, not to conflate or flatten culturally specific scholarly work but to highlight multiple fronts of critical scholarship seeking to dismantle oppression on a systemic level. Learning from these scholars, we can develop our *ecological community writing* work in systems as we imagine different and more just worlds.

A Local Examination: Assessing the Ecology Through Collaborative Research

Before my partners on The Shed: Boulder County Foodshed board and I did any work to educate the public about local food, for which there seemed to be professed public support, we needed to better understand the multiple reasons why the failure rate for local farms was so high—we needed a study of the *system* (Castle, "80 Percent"). We recognized that this information would directly impact what the public might need to know or be able to do to help resolve this problem. As one of our first collaborative projects, Brian Coppom, then-executive director of the Boulder County Farmers Markets and The Shed board member, proposed to me a project for a University of Colorado Boulder graduate student to investigate the high failure rate of small-scale, local food farms in Boulder County.

Brian's question was the impetus for a two-year cross-disciplinary set of projects between University of Colorado faculty, graduate students, undergraduates, and multiple external constituencies, which I detail over the next several chapters. The research project that I explain in this chapter is an example of mapping out an ecology—whom do we know; who can connect us with whom; who else is working to understand these issues? This research to study the ecology of actors working to make local food viable in Boulder County is a part of the *ecological community writing* methodology. Even if Rhetoric and Writing professors and students are not trained in social science or science, we can partner with others at our college or university to answer community partner questions with a potential outcome being a community literacy project or public dissemination of findings framed using communication strategies and public writing. The project models collaborative, interdisciplinary, trans-community *ecological community writing* for readers interested in launching similar initiatives or considering the variety of potential projects involved in an ecological approach to a community partner's question.

As I discussed in chapter 2, in *ecological community writing* we or our students can investigate local discourses through media discourse analysis, through interviews, and through canvassing. On an even more local level, it is generative to understand who is studying related issues across disciplines and programs at our university or college. To that end, I investigated who taught food-related courses and what food-related work was happening at the University of Colorado Boulder. After reaching out to the Office of Outreach and Engagement on campus and searching course listings, I had a preliminary list of professors and staff I wanted to meet. Many generative email exchanges and coffees with people in departments and offices from Mechanical Engineering and Anthropology to the Environmental Center and the Dining Services Office helped me to understand several of the people involved in food studies and local food advocacy on campus. With a goal of forming a cohort to work with The Shed on local food literacy projects, I believed we might consider a kind of cross-disciplinary *distributed definition building* in action—curricular and writing work and events inflected with the distributed local conversations, interests, priorities, people, and natural elements.

Because of mutual interest in the project as well as graduate student availability, I partnered with Dr. Peter (Pete) Newton in the Environmental Studies Program (whose research is on food systems in the United States and Brazil) and Dr. Phaedra Pezzullo in the Communication Department (whose research is on environmental communication and environmental justice). We received a $24,000 grant from CU's Office of Outreach and Engagement to develop the project that fellow The Shed board member Carl Castillo and I named Dig In! The grant partially funded Ashley Dancer, the graduate student in environmental studies recruited by Pete to research Brian Coppom's question of why the failure rate was so high for small-scale, organic vegetable farms in Boulder County. This research eventually developed into Ashley's master's thesis. It is from her master's thesis, *How Can Public Policy Create Opportunities and Barriers to the Development of a Local Food System? An Example of How Land-Use Policies Affect Small, Local Farmers in Boulder County*, that much of the information in the following paragraphs comes. Our collaborative, published article, "Perceived Opportunities for, and Barriers to, the Development of Local Food Systems: A Case Study from Boulder County, Colorado," expands on Ashley's initial research.

At the time of Brian Coppom's request in 2016, nineteen of the twenty-four small-scale organic farms that had opened since 2012 in the county had declared bankruptcy or shut down. Boulder County is not alone in its struggles to support small farm viability. In the United States, small family farms account for 90 percent of farms, and a majority fall in a "critical zone," defined by "a rate of return on assets of less than 1 percent and/or an operating profit margin of less than 10 percent" (Hoppe 5, qtd. in Dancer et al.). About 66 percent of small family farmers are in this critical zone for operating profit margin, and 75 percent are in this zone for rate of return on assets (Hoppe 6, qtd. in Dancer et al.). In other words, small-scale farmers in the United States are barely getting by, and many struggle financially. It is not surprising then that the country's number of new and beginning farmers is declining rapidly. This is coupled with the aging of established farmers; in 2012, over 31 percent of principal farmers were sixty-five or older, and the average age of farmers in 2012 was fifty-eight (Williamson, qtd. in Dancer et al.). But Boulder County has a strong record of supporting local agriculture. Why hasn't it been more successful, then, in keeping small farms viable?

Many residents of Boulder County value our Open Space. In fact, in 1967 Boulder became the first city in the United States where residents elected to tax themselves specifically for the acquisition, management, and maintenance of Open Space (City of Boulder). In 1977, within the first Boulder Valley Comprehensive Plan, Boulder County and the City of Boulder established the vision for Boulder Open Space, aiming for the "preservation of land for passive recreational use; preservation of agricultural uses and land suitable for agricultural production" (City of Boulder, "Boulder's Open Space"). Both the city and the county continue to support agriculture on Open Space through the Open Space tax, agricultural training opportunities, and the Boulder County and Open Space Cropland Policy (created in 2011 to define sustainable agriculture on county Open Space lands, updated in 2021). Given such support, it is surprising that a recent beginning farmer program aiming to encourage and support small-scale local organic farming with nearly $1 million in funding had close to an 80 percent failure rate, which is significantly higher than the national survival rate for beginning farm businesses of approximately 55 percent between 2007 and 2012 (Dancer et al.). As Brian Coppom and other members of The Shed agreed, this high failure rate warranted further examination into the possible causes that have made this model of farming such a challenge for farmers.

Pete and I served on Ashley's master's thesis committee, connecting her to partners at The Shed, who then helped to connect her to local farmers and other stakeholders for her interviews. Over almost a year, Ashley conducted semistructured interviews with farmers and ranchers who had leased or purchased agricultural land within Boulder County and farmed on that land within the past twenty years, as well as key stakeholders from organizations involved with the agricultural system in Boulder County, including people from Boulder County Open Space, City of Boulder Open Space, People's League for Action Now (PLAN-Boulder), which was involved in the creation of the first comprehensive plan and initial policies related to Boulder City and County Open Space, Colorado State University Extension, National Young Farmers Coalition, and Boulder County Farmers Markets. Through multiple events, farmers markets, and introductions from community partners and other farmers, we identified potential interviewees, and ultimately Ashley interviewed sixteen farmers (fifteen used organic practices, and one identified as a conventional GM sugar beet farmer) and six key stakeholders either in-person or over the phone at a time and date of the interviewee's choosing. The goal was to better understand the challenges and successes experienced by farmers in Boulder County. As Ashley reported out her research, I also paid attention for any need for future projects to address ideas or issues that arose from the research.

While Ashley's study revealed many benefits of farming in Boulder County as well as barriers to success, some of the key issues that emerged as barriers included the necessity for farmers to hold a second job or for a spouse who contributed to overall household income. As one farmer put it, they would "not be able to survive on farm income alone." Another farmer said, "Yes [we could survive] now, but it took ten years to get there." Anecdotally we heard from The Shed board members that most small-scale farmers in Boulder County (all but one farmer at the time of the study) have second or third jobs and cannot survive on farm income alone. This proved to be accurate according to Ashley's research.

Farmers also discussed a lack of knowledge sharing, especially from aging farmers. Ashley found that knowledge sharing is particularly important in terms of farm succession planning, the process whereby the traditions, skills, and other knowledge, as well as land are passed from one generation to the next (McCrostie and Taylor 5). The average age of principal farm operators in the United States in 2017 was 57.5, up 1.2 years from the 2012 census (US Census of Agriculture). As the United States farmer population continues to age, the need to successfully transition land ownership and farming knowledge from senior farmers to new and beginning farmers may be a top concern.

A possible future literacy project, suggested by one of the ranchers I connected with through my classes' field trips to her family's ranch (her son was a student in one of my classes) could involve the gathering and sharing of farmer and rancher stories. As she explained to me, ranchers possess a wealth of embodied knowledge that often isn't written down anywhere and that many want to share it so that it isn't lost.

Farmers in Ashley's study also mentioned issues such as difficulty accessing suitable farmland, lack of affordability, and high cost of labor, limits on the number of events they could hold on the farm, policy barriers to farm diversification, and poor soil quality on land available for lease, making it more difficult and more expensive to farm the land. Because of the high demand for Boulder County Open Space agricultural leases, any land that was available for a new tenant tended to be of lower quality, making it difficult for new farmers (or farmers who were seeking to expand) to find high-quality leasable land. Poor soil quality is an obvious concern for farmers because the quality of the soil contributes to the quality of their produce and how much they are able to grow. Some farmers and stakeholders mentioned an increased interest in carbon farming (also known as carbon sequestration), a term used to describe growing practices that suck carbon dioxide out of the atmosphere and convert it into carbon-based compounds in the soil that aid plant growth (Barth). Carbon sequestration tax incentives and farm succession programs would also help to keep agricultural land for agricultural uses.

In working to build a viable and just local food system, these interventions would help move us beyond the individualist rhetoric of "vote with your fork," which excludes large numbers of people from being able to purchase local food due to cost. The interventions also move us away from putting the burden on farmers to reduce costs of their products while they risk bankruptcy or navigate multiple jobs to stay afloat.

Even with a directed support program from the county, a predominantly economically stable and enthusiastic consumer base, and other factors that point in favor of a local food system, farmers in Boulder County found it hard to make small-scale local farming a viable business. As my students' research demonstrated, many community members believe that a local organic food system offers tremendous benefits. It seemed from Ashley's interviews with farmers, however, that there may be a disconnect with the general public about the expenses connected with growing food with organic practices. The benefits associated with local organic food systems may not match the price that community members think they should pay for food. This can mean that producers are forced into lowering the cost of their products and may not be compensated fully for the value they are generating. If communities and policymakers view local organic food systems as providing a benefit that cannot be fully internalized by the price of food products alone, there may be policy-level opportunities to provide support for producers that do not put an undue burden on either farmers or individual consumers.

Ashley's master's thesis original research became the foundation for her, Pete, and me to co-write an article about the perceived opportunities and barriers to farming in Boulder County and to share those findings with The Shed and other constituents in the county. The farmer and stakeholder responses spoke to concrete realities, both positive and negative, that shaped the viability of people's definitions for local food based on consumer understanding, environmental realities, and existing policy. We hoped that together we could work toward responsive public policy and food literacy projects like Dig In!

From Local Research to Local Action: Rhetorical Intervention

Ashley, Pete, and I shared our findings with the Boulder County Farmers Markets and The Shed board, who could in turn share the information with their respective organizations. Given the material realities—the century-old water rights laws that prevent farmers from being able to farm certain lands or access needed water; dated policies that restricted the number and kinds of structures farmers can have and the numbers of events farmers can host, such as farm dinners and wedding receptions, on their land; the short Colorado growing season; the intensive labor involved in organic production of diversified vegetables; the high cost of land in the county; and other factors—how might we as a collaborative partnership between the university and The Shed help consumers understand the complexities that inhibit small-scale farmer success? How might we expand our support of and public awareness about local food? These are the types of literacy questions that Rhetoric and Writing scholars and teachers are well equipped to address.

There are also many people who either cannot afford local organic produce or who remain unpersuaded as to its value. In my classes and at The Shed board meetings, we asked what programs and policies would need to be implemented or expanded to help make local food more affordable for both consumers and producers, as well as how we could connect public literacy about local food to literacy about systemic environmental issues, food access, and affordability for fellow community members. These are only a handful of the issues that my classes and The Shed board discussed as we decided how to best intervene in public discourse and policy deliberations around local food.

Given the nature of an ecological approach to community engagement and the distributed work happening already across the county to help define what is and is not feasible, what does and does not matter, and so on, I knew there were others who were better suited to tackle some of the issues the interviews uncovered, such as how to improve soil health or change policy to amend land use restrictions. For example, Pete is faculty in the University of Colorado Environmental Studies Program and the Masters of the Environment track in Sustainable Food Systems. We hoped that future courses or students would work with farmers on soil health and regenerative agriculture practices, and indeed several subsequent students' projects have worked on these issues with farmers across the county. adrienne maree brown's concept of emergent strategy is relevant here in thinking about working with multiple issues in a complex system. She offers advice to organizers that can be useful for community writing practitioners as we consider where and how to begin and the affordances of a systems approach through consideration of abundant knowledges and skills throughout the ecology. She explains,

> Emergent strategy is about how we shape and generate complex systems and patterns through our own relatively simple interactions. Nature moves in small fractals of interdependence, accumulating nonlinear changes and creating more possibilities with the constant adaptations of a resilient earth. If we attend to nature's lessons, we can remember that we, too, are nature; we can unveil our own organic gifts, our way to the future together, our path to thriving in this abundant world. (*Holding* 12–13)

brown so beautifully reminds us that change is nonlinear and fractal, distributed in small connections throughout a complex system. In our study of the barriers to and opportunities for local food production within the county, Ashley, Pete, and I took a systems approach, suggesting multiple leverage points where small policy changes could impact the system and where even small shifts in consumer literacy could do so, offering cascading effects that we hoped would increase farm viability. The change we hoped for based on the research would indeed be nonlinear and fractal, and in the next chapter, I describe several of the Dig In! projects that, through small or simple connections and actions, embody emergent strategy.

In thinking through writing for impact in complex systems, we can consider how to move people toward systemic rather than linear thinking. As David Peter Stroh delineates in his work on the many benefits of systems thinking, one benefit is most relevant in conjunction with brown's description of emergent strategy. Systems thinking "*focuses* people to work on a few key coordinated changes over time to achieve system-wide impacts that are significant and sustainable" (22). He urges us to consider Meadows's "leverage point" concept, a systems-thinking approach that "shows people their responsibility for current reality and uses leverage to change the few things that change everything else" (23).

Of course, I am only one part of any system, and it would be foolish for me or my students to claim sole credit for sizable and dramatic impact and change. As Kristen Seas explains, "Our field operates on a tacit belief in a direct causal relationship between our rhetorical practice (stimulus) and the change we see in an auditor (response). Such linear causality, however, is untenable from an ecological perspective because ecologies are, by definition, nonlinear systems in which 'small changes can have unforeseen consequences that ripple far beyond their immediate implications'" (Seas 52; Brooke 28, qtd. in Seas). Much like brown's concept of emergent strategy that focuses on small-scale actions that lead to larger, systemic changes—"Small is good, small is all. (The large is a reflection of the small.)"—Stroh and Seas highlight that there are an infinite number of micro- and macroshifts and impacts occurring continuously (brown, *Holding* 14). With these infinite possibilities of where to attempt impact, how might we think about where to begin to motivate change, catalyze collaboration, focus energy on key elements, and stimulate continuous learning in our on-the-ground projects?

On The Shed board, we wanted to encourage community members to take an active role in helping to determine how we (and board members' representative organizations) consider a broadly defined "local food" system. Community members' involvement in definition building would, in turn, help to drive cultural transformation. People's participation in CSAs, farmers markets, home gardens, food rescue, and the like materializes their embodied food literacy and, in turn, shifts the conversation about what people want and what is possible in the county as we define "local food." I hoped to take what we had learned through my students' and Ashley's research and create food literacy projects throughout our community to help inspire a locally significant and organically produced groundswell of support for local food through *distributed definition building*.

Rather than assume that the term "local food" reflects a fixed, already-determined reality or set number of miles, by using an *ecological community writing* approach, we can allow for multiple and perhaps competing conceptions of the term's meaning to emerge. By distilling the complexities of national food issues as well as the issues articulated by local farmers and stakeholders; by tapping into the connections and relationships highlighted in local media discourse and on nonprofit organization websites; and by considering the diverse responses given by community members during the canvassing my students had done at the farmers market, I recognized my role as a Rhetoric and Writing scholar-teacher was to encourage Boulder County community members to connect words to meaning and to construct their own definitions that would inspire the community toward literate action.

Theories of Contagion and Virality

In order to think about *ecological community writing* in action, perhaps it is helpful to think of the concept of local food as an idea that can be contagious. As Seas has written about rhetorical contagion, a community-engaged practitioner's work may be to help set a scene ripe for contagion. In "Writing Ecologies, Rhetorical Epidemics," Seas challenges us to ask, "[W]hat happens to determine the effectiveness of the text beyond the moment of composition," instead looking at "the larger life cycle of communication" (52). Any action I produce will have a nonlinear effect that I cannot anticipate. "Here," Seas explains, "effectiveness appears to be measured by a rhetoric's ongoing evolution, even as it mutates and no longer resembles any original intention or telos. Rhetoric as an art, therefore, is better understood not as an isolated exercise but both as a process of distributed emergence and as an ongoing circulation of process" (53). Seas's argument about rhetorical contagion influences my community-based writing work and is critical, I believe, to making our own writing and our students' public writing more impactful. Seas posits that the study of transmission of contagious rhetorics "can explain the spread of ideas and behaviors, [. . . and] can help us account for rhetorical effectiveness in network culture once we have abandoned the notion of intentional agency" (53–54). Taking Seas's cue, I turn to Malcolm Gladwell's *The Tipping Point* for inspiration. In my effort to help create a ripe environment for rhetorical contagion of the idea of local food, I draw on Gladwell's four interrelated factors that contribute to outbreak—the virus, hosts, environment, and time—and the view that "it depends entirely on how all four components together create the conditions for contagiousness and produce that vital but elusive tipping point when an idea takes off beyond all expectation" (Seas 55). Seas contends that "any rhetor must attend to and cultivate these conditions, rather than seeking to isolate specific strategies for controlling them through craft" (56). She explains, "These components must be seen as a whole ecological system that generates the conditions for effectiveness and makes it possible to cultivate not just cultural but *rhetorical* epidemics" (56).

How do community writing practitioners cultivate conditions ripe for rhetorical saturation? The first component, as Gladwell says, is that the virus (rhetorical frame) must be "sticky, . . . so memorable, in fact, that it can create change, that it can spur someone to action" (Gladwell 92). This rarely happens in one attempt. Rather, it requires ongoing revision based on feedback and repetition.

In the sociological theory of diffusion, with five social groups—innovators, early adopters, early majority, late majority, and laggards—Gladwell explains that

> [t]he biggest leap an idea must make, then, is from the early adopters to the early majority, whose acceptance then makes the innovation not only suddenly popular but eventually normative. Thus, this particular leap is the tipping point in the progression of the contagion, when the innovation is suddenly visible and widespread seemingly overnight, just like a biological epidemic. The key to overcoming this gap is communication. (qtd. in Seas 57)

Seas argues that early adopters are key in framing the idea in just the right way to make it seem important or exciting, which will make the early majority want to jump on board. The practice or thing is not what matters. She reveals, "[R]egardless of content, the communicable agent is actually the message that shapes the practice for potential adopters. What spreads is the rhetoric" (57). So, how the message is distilled leads to action. It isn't the local food itself, the thing we eat; it's the message *about* local food. It's the *rhetorical idea* of local food. What spreads isn't the thing itself; it's the frames! We must figure out what the current conditions are, the rhetorical ecology in which we're writing, and how we, as early adopters or partnering with early adopters, can help to tip it toward wider acceptance.

The next several stages of my *ecological community writing* methodology, then, involve determining current conditions and who/what are involved. Readers wanting to use this methodology can create scaffolded course assignments (that may span semesters or years), mentor students as they conduct public research, and facilitate connections across an ecology of actors who may or may not yet know one another. As I've previously described, in my own case of food literacy, this occurred through a series of projects that are all parts of the *ecological community writing* methodology: research about the issue at a national level, followed by my students' interviews with restaurateurs, discourse analyses of local media and nonprofit websites, canvassing at the farmers market, connections to faculty and staff across campus to create coalitions, interviews with farmers and stakeholders, conversations with The Shed board, and mapping of organizations and people in the county to find the innovators and early adopters.

Circulation of Ideas

When we try to make change, it does not happen in a linear fashion, though this, historically, is how public writing projects have often been structured, with a student or group of students producing a product for a partner. Rather, "given the velocity at which writing circulates, all facets of network societies are saturated by systems of writing, even those societies that do not directly engage writing but are, nevertheless, affected by writing's ubiquity" (Dobrin, *Postcomposition* 144). In these dynamic, moving systems, Dobrin posits "writing as a complex 'liquid' system" (179). He explains, "[S]aturation fills, makes a space full ... also suggests a sense of overwhelming ... or of completeness" (183). Dobrin is interested in "a posthuman, hypercirculatory concept of a system in which agents become indistinguishable from the system itself ... the system [is] the predominant feature of the ecology" (133). In other words, he believes that each agent is only as important as their place within the entire system. Gries similarly writes, "No longer does agency remain with individual agents; instead it travels, shifts, and evolves through the circulation of writing" ("Agential" 79). We, our students, and our project partners can study how a message has evolved as it circulates and what happens to it after the time of production.

Gries has argued that rhetorical life should be a focus of study, that there is still life with rhetoric beyond its initial production. "Rhetoric has 'life-spans' of its own," she explains, and "once unleashed into the world, it evolves, transforms, and transcends as it circulates and intra-acts with other human and nonhuman, concrete and abstract, entities" ("Agential" 79). Writing, then, is "a distributed process" and "may be dispersed across members of a social group, coordinated between internal and material and environmental structures, and distributed through time in the sense that earlier events can transform later events" (70). Given the distributed nature of writing, it is essential to think about these rhetorical "life-spans" as rhetoric circulates through communities. While it is difficult to study how the idea of local food moves through Boulder County to influence our collective consciousness, I can help create distributed writing opportunities and events that allow individuals to feel agency in defining what "local food" can mean in the county.

Community writing practitioners can help set the scene for distributed writing and study the circulation and remix of ideas in their ongoing transformation, some of which may have a powerful effect. We can try to write and help others to write for circulation. Rhetoric has a material force beyond the goals, intentions, and motivations of its producers, "more like an unfolding event—a distributed, material process of becomings" (Gries, *Still Life* 7). As we consider remix and circulation, we can encourage our students and partners to anticipate the "consequences" of writing in complex ecologies. For example, we can consider potential consequences of the writing for people with traditionally less power. As the concept of local food has traveled throughout the nation and Boulder County, specifically, my students and I can study the rhetorical roles or lives it has taken on. I can cultivate, and encourage my students and partners to cultivate, the affordances of our ecology to aim at individual and community-wide change. I distill these ecological and systems theories for students and project partners in ways that help them create impact in their writing and messaging. Rebecca Dingo argues that we must study "how rhetorics travel—how rhetorics might be picked up, how rhetorics might become networked with new and different arguments, and then how rhetorical meaning might shift and change as a result of these movements" (2). As rhetorics circulate, the connections they make can change their meaning. The writing of something like "local food's" definition is continuous, with endless linkages. In the next chapter, I will continue to isolate a few moments of this continuous, nonlinear, dynamic flow of *distributed definition building* in order to model the *ecological community writing* approach to developing curricula and partnerships.

Seeing the Elephant

I believed that The Shed board would be most successful in "tipping" the issue of local food if we could set a large-scale scene in Boulder County ripe for "rhetorical contagion" (Seas). We wanted to encourage the idea of local food's importance to "leap," so that the idea is not only popular or trendy "but eventually normative. Thus, this particular leap is the tipping point in the progression of the contagion, when the innovation is suddenly visible and widespread seemingly overnight, just like a biological epidemic" (57). An answer to how to "tip" and distribute an idea is at the heart of the *ecological community writing* methodology using *distributed definition building*—we (The Shed board, my students, and off- and on-campus partners) wanted the public to see the messages about local food in as many places and from as many angles and people as possible, and we aimed for saturation across the ecology, *which can simultaneously build the ecology*.

So as not to dissuade people from action, we framed messages through *critical rhetorics of abundance*, an abundance that does not shy away from the complexity of emotions, the complexity of policies, the complexity of historical realities of how agriculture began in this area, the complexities of agriculture's importance to the county and the local government's rhetorics about heritage. *Critical rhetorics of abundance* incorporate humans, organizations, things, ideas, language choices, and feelings as vibrant partners in creation. Then we engaged the community writ large to generate their own definitions. This *distributed definition building*, I hoped, would generate the most sustained impact.

While Ashley Dancer, Peter Newton, and I were involved in the research through the Environmental Studies Program, I also worked with a PhD student, Constance Gordon (now a professor at San Francisco State University), who studied food justice in CU Boulder's Department of Communication with Phaedra Pezzullo. We collaborated on a multifaceted food-literacy campaign to educate university students and the public about the local food movement. Constance was a research fellow funded through BoulderTalks (now called the Center for Communication and Democratic Engagement), run by Phaedra. Constance has published on food justice in Denver and elsewhere (see Gordon et al.). She, my undergraduate writing students, our partner The Shed, and I worked together on several projects proposed by The Shed:

1. The Shed needed help in redesigning their website to offer more information and resources to encourage citizen engagement and literate action. The website was static and hosted an unpopulated "Resources" link. They asked my students to use their knowledge gained during their readings and research projects in the first half of the semester to help create and curate resources, including blog posts for a new blog on their website, which was to be an ongoing collaboration each semester between University of Colorado Boulder writing students and The Shed.
2. The Shed board asked for help launching a nutrition campaign and food growth and preservation workshops to target lower-income people, working with the City of Boulder and community leaders to plan and implement workshops, educational demonstration gardens, and perhaps a food forest in a public park in Boulder to increase access to and production of free, healthy produce. They wanted University of Colorado Boulder students to be involved in proposal writing and workshop planning, execution, and follow-up.
3. The Shed board asked for University of Colorado Boulder students' help in launching and carrying out a campaign to educate younger community members, who may have influence over their parents' purchasing and growing choices.

As a community writing professor, I am inclined when engaged in these types of brainstorming conversations with partners to consider whether I can and should create a community-based course around these issues, with which students can engage through public writing projects. But clearly, based on even the preceding short list, the issues and partner requests are far too complex (and sometimes beyond my area of expertise) to be adequately approached in a single course. An *ecological community writing* approach across multiple courses, departments, and community members would be preferable.

Drawing on Gail E. Hawisher and colleagues' "cultural ecology of literacy," or literacy's embeddedness in particular contexts, Feigenbaum posits that "literacy *can* facilitate significant changes at the level of the individual and society but only in concert with a host of other factors that also influence the kinds of changes that occur" (Feigenbaum, *Collaborative* 31, original italics). The Shed's charge, then, as we saw it, was to encourage the policy and infrastructural changes necessary to better support local food production while simultaneously educating the community toward literate, embodied public discourse visible through action. Drawing on food justice literature, we knew that the actions available to community members had to be tied to more than just financial, "vote with your fork" options. We needed to think in terms of systems and ecologies. We drew on the abundant knowledges and skills available throughout our vast ecology of people working for a robust local food system.

- Several farmers and government leaders worked on the policy angle, leading farmer forums that resulted in successfully changing land use codes. The County, Parks and Open Space, and Colorado State University Extension conducted a 2018 study, and in December of 2018, the Board of County Commissioners approved amendments to the Land Use Code reflecting input from multiple studies and constituencies. In February 2021, the Boulder County Land Use Code was updated again to reflect farmer and stakeholder concerns, such as the number of allowed farm events, permitted structures such as greenhouses and hoop houses to extend the growing season, and changes to land use review processes (bouldercounty.org). In 2021–2022, multiple government agencies and CSU Extension conducted another study of Boulder County farmers and ranchers, leading to their significant report "Exploring Solutions to Boulder County Agricultural Sector Constraints," which outlines constraints and proposes solutions.
- Two of the county's innovative farmers, Kena Guttridge and Mark Guttridge at Ollin Farms in Longmont, launched Project 95 to work on soil health and regenerative agriculture projects in collaboration with Boulder County, community supporters, and young farmers "to lead a demonstration of what regenerative agriculture looks like on large agricultural Open Space parcels" (website).

- University of Colorado Boulder students in the INVST program worked with Micah Parkin, a Shed board member and executive director of the environmental organization 350 Colorado to generate a (sadly unsuccessful) plan for a food forest that would have been free and open to all on a city-owned property near one of Boulder's largest low-income communities. While I don't know what happened behind the scenes, the city decided instead to use the property for disk golf.
- Micah also created a Yards to Gardens program to help homeowners convert their yards to edible landscapes and worked to persuade the private owner of a large mobile home community to allow residents to grow produce rather than grass in shared areas.

These are only a few of the dozens of emergent projects as people across Boulder County worked to increase consumption of, production of, access to, and appreciation for healthy local food and agriculture.

Through my capacity as a writing professor, I took on a community food literacy piece to encourage an ecological approach to the problems of the definitional confusion around what "local food" is and can be, and to support The Shed's desire to encourage education about local food. Influenced by the ecopsychologists' and trauma scholars' conclusions that mobilizing change around emotionally complex social issues needs to involve storytelling and a feeling of community and collective action on a local level, I helped to frame several free, public events through a *critical rhetorics of abundance* lens to allow diverse members of the community to delve into some of the complexities around local food and create their own definitions based on their beliefs and values.

In the events we planned, I hoped we could harness the positive emotions people say that they associate with local food and use them to encourage more substantive inquiry and action on the personal level as well as at the level of community relationships, policy, and systems so that we might begin to address some of the barriers expressed by producers and stakeholders as well as customers. John R. Thompson argues, "'Eat local' . . . is a *call* for agency expressed in an action-based grammar readily available to the eater. 'Locavore,' therefore, is an identity based on empowerment and action" (67, original italics). To create our own "cultural ecology of literacy," The Shed and I launched our education campaign consisting of a variety of social media projects, public events, and a high school art competition. We wanted to call to action people of various age groups to think about what local food means to them. Using Thompson's concept of the power of an action-based grammar, we wanted Boulder County citizens to "Dig in!"—as in fork, shovel, inquiry—and then come up with a set of criteria that matters to them. We didn't want to impose a set of ethical criteria because those are far too subjective and contentious.

Even the members of The Shed board, all at the table because we cared deeply about local food issues, could not agree on what "local" should mean. What we could agree on, though, is that we wanted our community to know that "not all local food is created equal," as my students had argued in their farmers market project fliers. People cannot assume uniformity. What people can count on, though, is that because it is local, it is easier to determine farming practices, ingredients, and other factors important to them because they can ask or they can grow or see for themselves. That's a clear benefit of local food. Our distilled message then was: *Local Food: It's about more than just miles. Ask questions. Know Your Food. Dig In!* Because so many hundreds or thousands of people would be digging in, and because the digging in happens over time, this is not a linear causal model but an ecological one. Many things and people are acting simultaneously and continuously, leading to sometimes hoped for, sometimes disappointing, and sometimes unforeseen shifts, as Seas had said would happen with rhetorical contagion.

Part of ecological work is in understanding who is doing what in order to determine, to use Gladwell's terms, the innovators and early adopters as well as the potential early majority. Because of Gladwell's insight that the most significant leap an idea can make is "from the early adopters to the early majority, whose acceptance then makes the innovation not only suddenly popular but eventually normative," we would need to determine who the current leaders in the local food movement were and then the people we might encourage, through our carefully framed food literacy campaign, to become the early majority.

During one semester, I began informally mapping out on a large whiteboard the network of community actors with my writing students, as we considered which groups might have significant impact on "tipping" the concept of local food toward contagion. This wasn't a geographic map but rather a relational one with lines making connections between various organizations and people. We realized after an hour of this activity that the whiteboard was too small.

In the next class, I brought in large rolls of paper, laid them out covering the classroom floor and on tables for those who did not want to be down on the floor. I handed out markers to students, and they started creating a giant web of connections in what turned into an almost ridiculous, Twister-like game of moving around one another, crawling about on hands and knees, and connecting lines across the classroom. It wasn't the most polished pedagogical exercise, but it was a fun way for all of us to see how complex, intertwined, and vast the food-related work in the county is. A multimodal assignment related to this mapping can ask students to choose a mode for generating a visual representation of the ecology, be it digital mapping, knitting, poetry, quilting, board games, or other creative ways.

Mapping of Boulder County as an ecology of work and connections, human and other-than-human actors, inevitably embodies my own and my students' interests and speaks to what we know as well as what we don't know. As in any ecology, there is distributed, simultaneous work happening that each person individually cannot possibly know fully. This led to us to develop a series of questions about those gaps in our knowledge, and students either connected with members of The Shed or did research on their own to find answers. The Shed board members offered several leads (in fact, many of them were innovators in the ecology); my students helped fill in the relational map through their research into local projects and organizations, drawing, too, from the annotated bibliography of local food organizations created several semesters back by another of my classes. Sometimes the mapping of the ecology led to relationships—I would reach out to people to learn more about their work. Through our conversations, they might suggest I connect with other people I didn't know, or I would mention a project I had come across that might interest them. This process of connecting was rhetorical and generative. It also clearly indicated gaps in leadership and programming when it was visible who or what was missing from the map.

A parable from ancient Indian folklore, adapted by many cultures and often cited in systems theory, is the anecdote about an elephant surrounded by blindfolded people all feeling an isolated part of the elephant. No one knows they are touching an elephant. One feels the trunk and assumes it to be a snake; one feels a tusk and calls it a spear; another a leg and calls it a tree trunk; yet another the elephant's side and calls it a wall. In turn, each blindfolded person creates their own version of reality from their limited experience and perspective.

If our local ecology is the elephant, part of our work as community writing practitioners can be to make visible those connections between the parts, to help show people the elephant, the ecology, and to help them connect to other parts of it. And it's not just connecting to present parts. It is the history of the elephant—the systemic structures that have made the elephant what it is today. It is connecting all of us currently feeling the elephant and connecting to all of those before who have felt this elephant. Bernardo and Monberg's discussion of the "long arc" of reciprocity built into the Indigenous Filipinx concept of *kapwa*, "reciprocal being," explains, "When we think of kapwa or of resituating reciprocity within a framework of social justice or food justice, the arc of time expands deep into history and far into the future" (87). Maybe one of our jobs, then, as scholars, teachers, organizers, activists, writers, community members, and all of the many ways we define ourselves, is to help visualize or manifest the ecological nature of the work. In other words, our work doesn't just function as part of an ecology, although it does that as well. I am suggesting something different, that it is more than that. *Our work is ecological: it is a thread that connects.* Our work is a connection between scholarship, teaching, and service, between communities, between people, and beings, and land, and projects, and places, memories and emotions, and moments in history.

Our work is ecological. The affective elements are actors. The history of place is an actor. And all of us are writing, in the most capacious definition of that word, how in my county's case, "local food" can be defined. The other-than-human, affective, and human all circulate around each other, intra-act, are "vibrant matter," as Bennett calls it. "Local food," the concept, is vibrant, collaborative, distributed, and has "thing-power" (Bennett 2). Every seed, farmer, eater, and plot of soil has an ability to make things happen, to impact us and our communities, to help build locally responsive curricula; and to "write" change, as I will explore in chapters 5 and 6.

Maybe our challenge then is to help remove the blindfolds, even as we ask others to help remove our own. Maybe our work is to expose the elephant, not as a fixed thing to be seen and suddenly understood but as a continuously evolving, complex, and abundant system with writing as a thread that connects it all.

5

Ecological Community Writing in an Abundant Foodshed

with Constance Gordon

While my informal, relational "mappings" of contributors to the local food work in Boulder County (e.g., nonprofits, farms, government agencies) could not display ecological completeness, limited as they were by my, my partners', and my students' knowledge of who and what were working within the ecology, they did help to illuminate openings. They helped visualize where innovators could connect with potential early adopters and where tipping might occur. The mappings showed what there was and what wasn't yet, who had a voice and influence, and who did not. I began to consider the maps as verbs: as is and might be. Each of the actors has agency to co-write Boulder County's continuously dynamic, evolving definition of what constitutes "local food." Government agencies, farmers, media outlets, seed savers, chefs, nonprofits, and consumers were all writing that definition for what "local food" is here in this particular place.

As I discuss in chapter 1, according to Deborah Brandt's concept "sponsors of literacy" refers to "any agents, local or distant, concrete or abstract, who enable, support, teach, and model, as well as recruit, regulate, suppress, or withhold literacy—and gain advantage by it in some way... Sponsors set the terms for access to literacy and wield powerful incentives for compliance and loyalty" (166–67). As my students discovered through their interviews with restaurateurs, farmers, and farmers market patrons, as well as through media, when powerful sponsors promote a narrow definition of local food that may not align with consumers' understanding or desires, they are regulating and sometimes suppressing understanding. As a Rhetoric and Writing scholar, I am well positioned to intervene by building food literacy events that distribute power to define "local food" out to community members not often in positions of power.

This chapter highlights the range of activities my partners and I undertook using *ecological community writing* to foster multimodal methods that embody and communicate *critical rhetorics of abundance*. Through this community literacy work, we hoped to welcome in and inspire more and more community members to feel connection and involvement as "writers," as I saw them, of Boulder County's local food work through *distributed definition building*. In considering the climate communication and trauma scholarship, I did not want to frame events through threatening, apocalyptic, or fear-based rhetorics. We needed to bring joy and connection into the events even while teaching the important critical lessons.

In this chapter, I turn to the idea of distribution across human communities through many projects, all of which were aimed at helping people construct their own literate definition of local food and understand the systems they support and reject through their purchases and engagement. Through the events we created, The Shed board members and related nonprofits, my students, and I engaged in community food literacy work, providing ideas, resources, and materials, while we tried to allow for openness of responses as well. In thinking of the distributed and dynamic nature of rhetorical concepts and writing, we hoped to limit the constrictive elements of literacy sponsorship and encourage the more capacious possibilities of what is possible. At the same time, because of how difficult it is for farmers to thrive and for so many people to afford local food, we did not want our advocacy (as Gottlieb and Joshi so importantly remind us) to be "shorn of any social and cultural context" (183) and "benefiting primarily those who could afford such a diet" (183). A significant focus of The Shed's work was to support farmers so they wouldn't go bankrupt and to encourage the public to understand why developing local food production capacity is wise in the face of climate change and other environmental crises, not to mention to build community health in several ways. The Shed's goals were front and center as my students and I helped work on projects we co-created with The Shed. Part of the Dig In! goal was to foreground the conditions and limitations under which Boulder County farmers grow and harvest, sustainability or regenerative considerations along the production chain, and community members' ability to access and appreciate the value of locally grown food. A justice-oriented approach, we hoped, would encourage people to dig into the nuances and complexities while also examining their own values and how they might best access or grow food that aligns with those values. It also meant digging into the connected issues such as land use policies, gentrification and sprawl, housing costs, and ways in which people can distance themselves to whatever extent from an industrial food system that leads to environmental and human degradation.

To help students understand these systemic complexities and to bring in knowledges from as many people outside of the university as possible (always aiming to break barriers of where knowledge is housed, how it is shared, and whose knowledges matter), I invited and paid through Office of Outreach and Engagement grants several of my community partners and nonaffiliated community activists to facilitate discussions with my students about their key areas of focus. My students and I toured several locations in the county, as well, from the Community Food Share, a large food bank in Louisville, Colorado; to farms, dairies, and ranches; to Community Table Kitchen, which trains and provides jobs in the service industry for unhoused people in Bridge House's Ready to Work program. Through all of these visits in and outside the classroom, my students and I discussed ecologies of local food and engaged in the relational mapping exercises I discussed previously.

As I've continued to read and learn over the years since the Dig In! projects began, especially from activists and colleagues of color, I would frame my definition of "justice" differently now and think about the ways in which whiteness and privilege were showing up at each event. At the time, I was thinking about justice and my conception of "critical" in a less robust way, focusing primarily on food access for community members; small-farmer livelihood and farm success; and the environmental, relational, and economic benefits to communities that try to disconnect, to whatever extent, from the industrial food system. All of these considerations are extremely important, but there is more to consider.

Critical Goals

Because my ideas are always changing (I hope in generative ways), I'd like to approach this chapter through a lens of learning and curiosity, as my ideas for what an *ecological community writing* methodology can offer are always developing in terms of both how to catalyze *distributed definition building* and how to ask challenging questions about the efficacy of my work, as well as to whom or what I hold myself accountable. Nothing in community-engaged writing work is static. Therefore, I offer methods for implementing an *ecological community writing* methodology while also offering questions along the way for how I might better have achieved justice-focused, "critical" goals. I hope that this discussion will help you to imagine ways to create projects and events to connect and grow the ecology of work happening in your own area of focus.

In service-learning scholarship, Tania Mitchell offers community-engaged teachers an important distinction between traditional service-learning and what she calls critical service-learning. The definition of community-engaged writing that I used when I first developed the Writing Initiative for Service and Engagement (WISE) at University of Colorado Boulder is that it is a form of experiential education that integrates academic instruction and regularly scheduled critical reflection with educationally meaningful community-based work that is appropriate to curricular goals and is meaningful for the community partner to whom we are accountable. Students' participation in this writing enriches and enhances the learning experience, teaches how to become involved in community efforts through writing and rhetoric, and meets community-defined needs (adapted from the National Commission on Service-Learning). This definition deepened over time thanks to Mitchell's and others' insights.

While students and community members may be making connections between readings, research, and community-based observations and activities, and all of this shows learning, they are not necessarily performing work with a "goal, ultimately, . . . to deconstruct systems of power so the need for service and the inequalities that create and sustain them are dismantled" (Mitchell 50). This is a distinction between traditional service-learning and critical service-learning. Mitchell explains, "[A] critical service-learning approach allows students to become aware of the systemic and institutionalized nature of oppression" (54). Following Mitchell's lead,

I ask, "How might the curriculum, experiences, and outcomes of critical [*ecological community writing* courses and projects] differ from traditional [community writing courses and projects]" (50). In my examination of the projects that I helped facilitate, I want to keep coming back to this question.

In a similar vein, we can return to the distinctions alive in food justice activism that I discussed in chapter 1. In the "Introduction" to their collection *Black Food Matters*, Hanna Garth and Ashanté M. Reese explain, "[A]t minimum food justice highlights the problem of the contemporary food system that relies heavily on undervalued labor and the quick and efficient circulation of food products and is concerned with the unequal distribution of and access to healthy, affordable food" (Garth and Reese 1). I believe that the work I have been involved with hits this marker for working on food justice. However, Garth and Reese explain the need to push the work further:

> [M]any Black activists and scholars theorize and frame food inequities as a byproduct of the contestation to Black life, specifically grounding their food justice work from a starting point that just as racism produces increasing surveillance, disproportionate rates of mass incarceration, and income inequities, it also produces food and inequities.... The implications for those of us who study food is that it is not enough to simply examine race in the food system. We must also consider how the food system is part of larger structures that, by design, were never created for Black survival. (1)

This discussion of the interlocking, racist systems of which the food system is part highlights the importance of taking a systems approach rather than an individual manifestation or individual approach to the subject *and* to incorporate sovereignty and survival into our work. As I've shown throughout the book, this argument about Black food justice work is culturally unique *and also* related as well to settler colonialism and critical food sovereignty work being done by Indigenous, Latinx, and immigrant and refugee activists, organizers, and scholars. Racist systems can also be ableist systems, can also be settler colonial systems, and so on. As Mitchell argued for service-learning practitioners, so too do people working on food justice often distinguish between the immediate actions (e.g., the events I describe that get community members energized about and skilled in some of the important elements of local food production) and actions that get at some of the fundamental problems in our food systems, problems that include connections between race, settler colonialism, systemic issues, and the legacies of historical trauma tied to current practices.

To be "critical," *ecological community writing* pairs an examination of these systemic issues with action. In offering discussion of the projects we facilitated that had elements that worked well and elements that did not work as "critically," I can name where I was employing the methodology but without centering the work that Garth and Reese would say makes it truly focused on a robust definition of justice. I am continuously learning and developing a definition of justice in my work while humbly recognizing my own long road ahead. The work is a continual process, not a destination, and so I approach my previous work with curiosity and an openness to continual change.

Building the Ecology Through Saturation and Distributed Writing

The confusion over how to define "local food" in Boulder County offered a *kairotic* opportunity for collaborative and distributed public writing. The Shed board hoped to encourage the many policy and infrastructural changes necessary for the local food supply to meet demand and farmer needs, while simultaneously educating the community toward literate, embodied public discourse visible through literate action. It was important to include people in these conversations who are typically denied definition-building power. I wondered how power over definition building might be better distributed to people of multiple ages, cultures, and backgrounds, including the roughly 13 percent of the Boulder County population who are food insecure. I also wondered how I, through my capacity as a Writing and Rhetoric professor, could encourage an ecological approach to the varying definitions of what "local food" is and might be, as well as support The Shed's desire to illuminate rather than obfuscate complexities associated with support of a local food system and Boulder County farmers.

I suggested that we use *critical rhetorics of abundance* to frame communication about the profound connection and gratitude people can associate with local food, community building, and connection to the earth. Affective abundance framed our educational project, Dig In!—a food literacy project that encouraged activities and events that allowed people to dig into information about industrial and local food and to learn skills for resilience (such as how to save seeds and how to grow, can, preserve, and access local foods) while experiencing community and a feeling of agency that ecopsychologists believe are vital for substantive engagement. Because there are such conflicting definitions about what "local food" is, I suggested to The Shed, my students, and my on-campus partners that we create opportunities for community members to write their definitions through inquiry and engagement that center connection.

Through a relational methodology of *ecological community writing*, we attempted to saturate the community with messages about local food. We offered distributed writing opportunities for more than a thousand community members to have the ability to shape a definition of local food that makes sense to them. Climate communicators have suggested that we must allow for feelings of loss, nostalgia, guilt, grief, joy, and collective pride. I set out to find ways to combine critical inquiry and affective engagement that allowed for the abundance of knowledges, skills, and definitions and for the emotional complexity necessary to consider issues systemically.

While we cannot know what ideas will catch on, we can work toward rhetorical contagion. In *Contagious: Why Things Catch On*, Jonah Berger posits the importance of creating a conversation piece. He explains social epidemics as "instances where products, ideas, and behaviors diffuse through a population. They start with a small set of individuals or organizations and spread, often from person to person" (4). The Shed hoped that if the board and affiliate groups acted as the connectors helping to create ties between groups and individuals who might not otherwise know one another and inviting people outside of The Shed to participate in our events, these "weak ties then connect clusters of strong ties . . . to other clusters that they would otherwise not have access to . . . these connectors might link an early adopter to a cluster of early majority types . . . the weak ties embodied in these connectors allow for precisely the kind of unexpected nonlinear jump required to trigger an outbreak" (Seas 61). Part of my role on The Shed board, as I saw it, and part of the role of any Rhetoric and Writing practitioner involved in community projects to increase awareness and capacity, can be as a translator of sorts, bringing the findings of studies and scholarship, in my case from multiple disciplines, into concrete actions.

In this instance, I drew from scholarship from ecopsychology, trauma studies, communication, systems theory, contagion theory, and so on to craft the events and projects I discuss in this chapter in order to meet The Shed's goal of "increased production, consumption, and appreciation of local food" (website). Through frequent and multiple projects throughout the county, we hoped to encourage rhetorical contagion, or the uptake of the idea of local food across the ecology by what I think of as saturating the community (through the method of *distributed definition building* opportunities) with the idea in as many places and through as many events as we could. These events would act as the conversation pieces Berger suggested would bring people together.

Ecological community writing offers a methodology for working toward saturating an ecology with an idea. I used the concept of distribution to create projects with several different partners and audiences to consider how to saturate the community with ideas and feelings about local food through events, projects, and communication that would help to *generate an ecology* of local food work and awareness. Many, many projects were happening simultaneously on many fronts and with many audiences, all with a goal to spread awareness across the community. The work was not linear, and I ask you to imagine a three-dimensional image of a continually expanding formation of connections that my, my students', and my partners' work was generating. If "local food" is at the center, the connections extend to parts of the ecology such as language used nationally, locally, and in media; affect and emotion (individual and collective); companies and organizations; epistemologies and other disciplines; humans (many of whose voices had previously been ignored); and other-than-humans. As I've shown through each chapter, our work as community writing practitioners can be to study and create opportunities for those connections to build or strengthen through a wide variety of writing projects.

Building an ecology requires identifying gaps. At a Shed meeting, fellow board members and I sat around a conference table considering current barriers according to Ashley Dancer's research and other testimonies. We discussed which groups of people we wanted to reach out to that we were missing in our current communication and how we could make more connections. Through this identification, we developed ideas for projects that my students or the represented organizations on The Shed board could tackle. I share several of them in the next few pages as examples of working toward saturation and distributed power of definition building by community members.

It helps to think not only of the connections in an ecology but of the unconnected parts that our work can help connect. The following project examples show ways we can think of projects saturating multiple parts of a community with multimodal messaging and opportunities for engagement.

WEBSITE AND BLOG WRITING

A clear problem emerged in discussions with The Shed. While there was quite a bit of work for and support of local food production and distribution in the county, people often did not know of the full scope of projects and organizations working on food-related issues. The gaps in knowledge would sometimes lead to replication of work or lack of resource sharing.

The Shed wanted a dynamic website that would encourage people to learn about local food in an ongoing way. The webpage was currently static. We realized there was not yet a comprehensive public online resource about local food in Boulder County that could provide answers to questions such as: Which farms provide CSAs? Which kinds of produce and meats are available each month? Which organizations offer gardening and canning classes? Which food events are occurring? What are other parts of the state and country doing in terms of programs and policy to impact local food production and consumption that Boulder County might adopt? We wanted to make it easy for interested people to know where they could go to learn more, connect with others, and promote their own work.

I suggested to fellow board members that our website house this information on a resources page and that my writing students could research local food organizations, farms, and companies and help to curate content for The Shed website. In a public writing assignment, groups of my students wrote to many of the agencies they and past semesters of students had found to not only inform them about The Shed's mission and website goals but also to ask whether they would like to add information to the resources page of the site. My students' work created ecological threads between past work, current work, and people in our community. Their writing helped build a definition for "local food" through what they included and omitted as important.

Over multiple semesters, my students also wrote blog posts for a new section of The Shed's website, *The Scoop* blog, that we developed specifically for CU students, faculty, and staff to contribute to, indicating a mutual commitment to ongoing partnership. In my classes, students considered the public audience that would be drawn to this blog, how to use rhetorical strategies we had studied to persuade, motivate, invite, and educate, and they played with visual and written rhetoric to create posts to spark community interest. Topics varied widely, from "Benefits of Eating Wild Local Game" to details about the newly implemented "Sugar-Sweetened Beverage Tax in Boulder County" to "Human Trafficking Among Sheepherders in Western Colorado" to "Is Marin County, CA's Carbon Project a Model for Boulder County?," and more.

We created a shared Google Doc consisting of blog posts chosen from over 300 pages of student-generated material. The inquiry topics were chosen by my students based on their own food-related interests as well as information we felt would be most useful for Boulder County community members. We gave The Shed's social media manager full access to schedule multiple posts per week. If a student wished, their name would appear in the byline, or they could remain anonymous. These posts were circulated across The Shed's and The Shed's partners' social media platforms as well. As students added to the blog each semester, they could read what previous students had researched, learn from them, cite them, and see that their own work became a circulated part of the dynamic ecology of local food literacy. Their writing helped to build ecological connections, and definition-building power was distributed across students who would typically not have a voice in influencing public discussions around how to use language related to a social issue.

MULTIMODAL WRITING

In two other assignments that help illustrate the saturation model and asked students to consider how to circulate ideas and messages, some student groups chose to make videos requested by local farmers and a nonprofit; these videos could circulate across social media platforms and could be used by the farmers and nonprofit on their own websites.

In another assignment, The Shed requested that my students create a bumper sticker slogan. From several slogans The Shed selected, we made the bumper stickers with Dig In! grant money. Once the stickers were distributed at our events, we watched them literally circulate through the city as cars moved the ideas and messages around the streets of Boulder County. It was a wonderful exercise in genre to have students consider how to best distill learnings from their research papers into public blog posts and then to distill the essential messages of the blog posts into succinct bumper sticker messages such as "Local Food: It's About More than Just Miles." The challenge was to make a pithy phrase that encouraged nuanced thinking rather than simple soundbites.

PUBLIC PRESENTATIONS

My students wanted to spread their awareness on campus as well. They reached out to professors who taught food-related classes in other departments, and several students asked to give a short guest presentation in those food classes on what they'd learned about local food studies in Boulder County. At least a half-dozen student groups from my classes participated in these cross-campus food literacy opportunities. *Distributed definition building* happened through the information my students shared with others and, we hoped, through the ripple effect of those people hearing the information, forming their own literate ideas of what "local food" means to them, and then sharing that information with others on and beyond campus.

SKILLS WORKSHOPS AS LITERACY WORK

A goal in our food literacy work with The Shed was to give ways for people to enjoy the abundance of produce when it is in season through teaching practical skills like food preservation to reduce food waste. We save it, and in saving, we can also share and trade, perhaps for another skill we don't know how to perform. As with recognition of the multiple knowledges that surround us, there are multiple skills we can choose to share. In another move toward saturation, I and several partners organized skills workshops to encourage canning and food preservation techniques so the produce could be enjoyed and distributed across community members after the harvest ends.

Through the Dig In! grant, I purchased materials and supplies for multiple events with one of my community partners, Micah Parkin—a Shed board member, executive director of the climate change organization 350 Colorado, and founder of Community Fruit Rescue, which partners with homeowners to harvest fruit trees or to forage wild trees in the city limits that have fruit that would otherwise fall to the ground and go to waste. They divide the bounty between homeowners, volunteers, Boulder Food Rescue, and other food access organizations to distribute to people who want fresh fruit. The 2022 harvest season totals listed on their website include 22,964 pounds of fruit harvested in 134 harvest sites. Food rescue, more broadly, can also be called food recovery or surplus food redistribution, which involves taking food that would go into a landfill from farms, grocery stores, and restaurants and redistributing it in communities that can use it. For example, Boulder Food Rescue's homepage reads, "Redistributing produce and power. Join us in creating a more just and less wasteful food system." Boulder Food Rescue runs several No Cost Groceries located at low-income housing sites, senior centers, or at after school programs and preschools. These No Cost Groceries are run by residents who live at the sites or in the immediate community. The organization's 2021 annual report says that 498,500 pounds of food were recovered that year; 963,259 pounds of CO_2 were diverted out of landfills; and 6,000 participants were reached. Through Boulder Food Rescue's and Community Fruit Rescue's work, they encourage waste and distribution to be considered in a definition for what local food can and should be.

With several of these organizations, Micah and I organized free fruit canning and preservation workshops for the community, which took place at public schools, places of worship, a public library, and public parks, to encourage widespread attendance distributed across the county. Again, we were considering an ecology of people and how to connect to them through these events. At these events, we distributed educational materials my students wrote. In framing the serious issues of food insecurity, food access, and food waste alongside ideas of bounty, preservation and sharing of the harvest, and community connection and relationship building, we offered food literacy skills framed through *critical rhetorics of abundance*.

The school events with 350 Colorado and Community Fruit Rescue taught parents and children about food waste and hunger in our community while, as figure 5.1 shows, they learned how to can local apples and pears. At the canning events at elementary schools, children took turns stirring big vats of apple sauce that they would carefully spoon into jars. The excitement of turning fruit into a favorite snack was written across their faces. They and their parents became part of the local food ecology.

At the Fruit Rescue Harvest Festival later that month, Community Fruit Rescue shared with attendees the volunteer portion of the fruit that was gathered, and canned it for later in the year when local food would be scarce; there was enough to go around. From young to old, as figure 5.2 shows, people smiled as they strained to press the apples in the cider press, knowing they were about to share cups of the sweet cider with the community that

FIGURE 5.1. Micah Parkin leads a canning workshop at a local public elementary school. Workshop supplies were purchased through the Dig In! grant. (Photo credit: Veronica House)

FIGURE 5.2. Woman uses a cider press with apples rescued from trees around the city at the Boulder Fruit Rescue Harvest Festival in a public park. Some supplies for the event were purchased through the Dig In! grant. (Photo credit: Veronica House)

gathered. Each person went home with apple butter, pear jam, apple sauce, and other treats that would keep through the winter. They all became part of the ecology, and through their embodied engagement, helped to write a definition for what is possible and important to consider in a definition of "local food" in the county.

The frame of critical abundance for these events can perhaps be understood through the lens of Kimmerer's discussion of the gift economy in her article, "The Serviceberry: An Economy of Abundance." She describes accepting the gift of berries while they are ripe and preserving some to eat herself and some to give to friends and neighbors. Kimmerer explains,

> Gratitude creates a sense of abundance, the knowing that you have what you need. In that climate of sufficiency, our hunger for more abates and we take only what we need, in respect for the generosity of the giver... To name the world as gift is to feel one's membership in the web of reciprocity. It makes you happy—and it makes you accountable.

Key elements of *critical rhetorics of abundance* are modeled in her story.

Framing the events we hosted through *critical rhetorics of abundance*, then, was not about hoarding and stockpiling. It was relational. We also wanted to discourage overwhelm in participants—it was not about thinking that one person alone must learn all the skills and hold all the knowledge. These can be distributed throughout the community, and there is a seasonal rhythm to when we have more and when we have less, when one person offers their skills and then steps back to let another person offer theirs. When communities take a relational approach to growing and distributing food, there is something quite beautiful in knowing when we have enough. I see this kind of community work as actively battling a pervasive focus on accumulation and unlimited growth, on always wanting more while promoting ideas of scarcity and individualism. Using *critical rhetorics of abundance* in these instances of skilling classes and distribution of food responds to this cultural sickness by opening capacity for relationships and collaborative community-building outside of the industrial food system and other systems of oppression.

Drawing upon ecopsychology's emphasis on affect and collaboration as key to moving people toward action, we wanted to help shape the ways people know and *feel* about local food, that it is political, it embodies forms of justice or injustice, it is accessible or not, it aligns with their ethics or does not, and it can be an abundant and joyful gift that brings people together in community.

As my students and I discuss every semester, there are all sorts of contradictions in our attempts to eat "correctly." Quantity-defined abundance has its limits and is paradoxically connected to food insecurity for others, as we learned through a Boulder County Farmers Market study that found that "the entire season's sales for the Boulder and Longmont, CO [farmers markets] could feed [the Boulder County population of 322,000 people] for only . . . a day and a half" (Coppom, "Interview with Boulder County" 81). Part of a critical frame in thinking through using abundance rhetoric in discussions of local food is in recognizing limits and distributing the harvest more effectively and ethically throughout our communities with equity in mind. This acceptance of limits or "enough" may seem contradictory in discussions of abundance, but it is not. It actually encourages a shift from an accumulation/hoarding ideology to an ideology of living with community in mind. We are all part of the ecology. The connections between us and the affective abundance coming through those connections can spur radical imagination of what we can create together.

SEED SAVING WORKSHOPS AS LITERACY WORK

Nestled in the heart of North Boulder is a five-acre property, Dharma's Garden, which feeds more than a thousand people in the community each season and provides a bounty of produce for my family every year. As I will discuss in more detail in the next chapter, I have partnered with the founding farmers, Tim Francis and Kerry Francis, on numerous events they have held on the farm to educate the community through frames of joy and abundance. Education about our connection to food and to the earth is an essential part of their mission. My writing students have spent several weekend days there over the years on education and work trips, and my Dig In! grant helped provide materials for Dharma's Garden vegetable fermentation and seed saving workshops, which were open to the community.

At one multiday seed saving workshop, participants delved into histories of seed saving in Indigenous cultures, contemporary work to preserve heritage seeds in the face of multinational seed companies, discussion of plants native to Colorado, and foraging laws. As we stomped on seeds to separate them from the chaff during one activity, Kerry and Tim broke out in dance. As we laughed and cheered, I thought of how many generations

FIGURE 5.3. MASA Seed Project founder Richard Pecoraro leads a seed saving workshop at Dharma's Garden. Participants work together in the seed saving process. Supplies were purchased through the Dig In! grant. (Photo credit: Veronica House)

of humans may have danced to save seed, have gratefully saved for the next season what has been offered from the abundant earth. That spontaneous dance—bodies moving in connection and gratitude to the earth, the seeds, and the nourishment that will feed us—is the epitome of an embodied *critical rhetorics of abundance* and helps to build a definition for what "local food" means in Boulder County. The seed saving event pictured in figure 5.3, the seeds themselves, and the people involved became part of the local food ecology. In each of these instances, farms, people, affect, and public spaces connect through the events—the work creates connections; it builds and deepens the ecology. *Our work is ecological.*

FACULTY COHORTS

As these projects with The Shed and connections across the community deepened and became more numerous, I proposed and co-led with my writing program colleague Paula Wenger a year-and-a-half-long cohort of five professional writing, technical communication, and business writing faculty. They received a departmental stipend for creating writing courses with at least one unit relating to community-engaged local food issues. In the first year of the cohort, we read articles for discussion on ecological theories of writing and community-engaged course design, and I shared with them about The Shed and projects about local food in Boulder County. In the third semester, I functioned as a community partner of sorts, a representative of The Shed. I visited each colleague's class and led discussions with students about current food issues in the county with examples of how previous writing students' research had been used. My colleagues' students and I brainstormed their own potential public writing projects and educational events for other students on campus, which they developed with their respective professors. They presented drafted projects to me for commentary before implementing them in the final weeks of their courses. My colleagues and their students then became part of the ecology of people writing with a goal of encouraging local food literacy.

Through all of the projects and events I worked on, both on and off campus, the goal was to saturate the campus and broader community with offerings that would resonate and give people a sense of connection and investment in our local food system, while educating them on what it means to support local food producers and think more deeply about the individual, community, and systemic implications of foods they eat. For community writing practitioners, all this discussion leads to questions of how we can approach writing for, with, and as communities to be most effective. Through The Shed events and affiliated workshops, we offered hands-on community building and coalitional engagement to rouse emotional connection, strengthen group bonds, allow people to engage with shared values around environmental resilience, and support farmers and food-insecure community members. All the while, participants learned important information that might motivate and energize them toward further action and circulation of information. These intended impacts, we hoped, would allow for people to feel they were acting and doing amid learning.

PUBLIC ART FOR DISTRIBUTED DEFINITION BUILDING

For the project and event I describe next, I was guided by the scholars across multiple disciplines whose work I've explored in the previous chapters. I discuss it in more detail than the other events described earlier in order to offer an example of a public enactment of *ecological community writing* through *distributed definition building* in sometimes useful and sometimes problematic ways. I will describe the scope of the project and the resulting event and will then look at it and the other Dig In! projects more critically to determine how my students, partners, and I might have elicited more critical, justice-centered learning from participants. I'll also come back to my own learning and how it has impacted the *ecological community writing* methodology.

When some of The Shed board members would visit my classes to talk with my writing students about where points of concern or interest were in their local food goals, we would talk as a group about whom to connect with next. This was an ongoing ecological and systems thinking discussion looking at possible points of impact and people to involve. While different audiences would emerge each semester, my students continually voiced their wish that they had developed their own food literacy earlier in their lives. They wanted to connect with other teens.

Over two semesters, my writing students, The Shed board, and I, along with then communication PhD student Constance Gordon, partnered to work with one population we hoped would serve as connectors to other parts of the population—high school students. The Shed board member Ann Cooper, who was director of the Chef Ann Foundation, and at the time of the project director of Food Services for Boulder Valley Public Schools, already had deep connections to all of the schools in the district through her trailblazing work building a national model for healthy school food programs. Ann connected me and Constance with high school art teachers at all district schools to generate our project together.

We determined to start with ninth- through twelfth-grade students specifically because we hoped not only that we would inspire teens but that their literacies would be discussed with their teachers and parents, leading to, perhaps, shifts in perspectives and purchasing decisions in multiple groups in the ecology. Constance and I connected with all high school art teachers in the district through email, in which we introduced ourselves and the larger food literacy project to ask whether they would like to build local food literacy into an art unit. Once we saw that there was enthusiasm, we developed a high school art competition through the Dig In! project, a county-wide competition geared toward high school students, asking them to creatively depict what local food means to them.

Through Dig In! grant money, Constance and I held an event at the Boulder Public Library, catered with locally sourced foods from the Farmers Market's Seeds Café, for the school district's high school art teachers and some farmers who expressed interest in participating in this project. We talked with the teachers about why we were proposing the art competition and led a discussion with them about possible ways to incorporate local food issues into an art unit that would allow students to think critically about the environmental, community, and social justice issues related to local food and to dig beneath the surface. Constance created a teacher packet and presented "Digging Deeper: A Food Justice Perspective" and "Utilizing Art for Social Change." Those teachers interested in working with us would receive a stipend as well as funding for all art supplies and associated production costs for their classes' artwork, paid for through the Dig In! grant. High school students would learn about local food and art's connection to social change, dig into one area they found most interesting, and creatively depict what local food means to them through art. They would write a reflection on the piece, and the art and reflections would be displayed at an art show at a local museum. Writing and art would become the catalysts for further questioning and articulation of issues. The high school students would be involved in *distributed definition building* and would help us build the ecology of local food work through their connections in the community to friends, relatives, and teachers.

By circulating some of the most compelling pieces from the competition around the county on social media and through future events, we hoped that local food literacy and interest would grow. Laurie Gries tells us, "As a distributed process, cognition may be dispersed across members of a social group, coordinated between internal and material and environmental structures, and distributed through time in the sense that earlier events can transform later events" ("Agential" 70). The art competition and other events would also drive people to The Shed's website, which my former students had populated with accessible resources and information for the general population; that way, people could formulate their own definitions of what "local food" means to them based on the issues most important to them. Then they could decide whether to buy, grow, or vote accordingly.

In collaboration with The Shed's Education and Promotion Committee that I chaired, Constance and I created a Dig In! Art Contest prompt, which guided the high schoolers in their investigations and art. The following is the prompt we collaboratively produced for the art competition.

> *Not All Local Food Is Created Equal*
> The *Dig In! to Local Food* Art Competition: *It's About More than Just Miles*
>
> "Local food" can be a challenging idea to define or bring to life beyond soundbites. It is often understood as simply food that hasn't traveled far to our plates. But the term "local food" isn't just about distance. It is about community and connection; it is about transparency. Being able to ask our local farmers and producers about how they grow their crops, treat their animals and workers, and produce their products allows us to connect with our food. We believe that local food conversations should also include our farmers' and farmworkers' ability to make a living and feel supported by their community, as well as healthy food access for all members of our community, regardless of race or ethnic background, physical and mental ability, or income level. Just because something is "local" doesn't mean it is necessarily any of these things. The power of "local" is in our ability to ask questions, or better yet, to grow or go see for ourselves.
>
> What is most meaningful or important to you or your family about "local food?" It's time to Dig In! and get to know your food by participating in the *Dig In! It's About More than Just Miles* art competition, hosted by The Shed Boulder County, a regional coalition supporting our local foodshed and economy.

We challenge you to become knowledgeable about what you are putting in your body, where it comes from, and ways to support community-building efforts.

Not all local food is created equal. Just because a product advertises that it is local does not necessarily mean that it carries the values you wish to support. **So, ask questions and dig in!**

1. Think of your favorite foods, then find one that is produced locally.
2. Dig into it (or its ingredients) to learn as much as you can about it. This may mean asking questions at the farmers market, or may require a visit to the local farm, garden, or kitchen where it originated.
3. Ask one or more people with knowledge about where that food came from how and by whom it was grown, raised, harvested, processed, packaged, and transported before it went into your mouth. Think about the impacts of all those stages and about the complexities around this food.
4. Was anything unexpected or different than what you had assumed when you first chose the food in step 1?
5. Can you dig deeper to consider how this food ties to bigger issues such as those listed above?
6. Given all that you have discovered, creatively communicate "local food's" meaning to you through words and art.

Constance arranged field trips for the high school students and teachers to visit local farms and ranches to help with their research and allow them to take photos and notes for their artwork and to ask questions on site to deepen their understanding of issues (see figures 5.4–5.5). As with the field trips I arrange for my writing classes, we also wanted them to begin to see the ecology of local food work—of people, farms, dairies, butcheries, food banks, rescue organizations, schools, and all of the other places they visited. That embodied visualization of the ecology through going physically to several places, we hoped, would start to build more of an ecological perspective.

Constance and I visited several of the high school art classes to offer feedback on the students' projects-in-progress and to talk through questions students had. In the first iterations, for example, one student created a beautiful painting meant to be of local fruit but included an orange in the painting. As oranges are not local to Colorado, we discussed options. The student could

FIGURE 5.4. High school art student observes cows for her artistic representation of local meat production. (Photo credit: Constance Gordon)

FIGURE 5.5. High school art students tour a greenhouse. One student is taking photographs for her art project. (Photo credit: Constance Gordon)

paint the orange out of the piece, or she could discuss in her written reflection the learning about what is not available locally and what this would mean in terms of eating seasonally and omitting certain foods if she were to attempt a local food diet. She chose the latter, as she felt it would allow for a richer discussion of the choices that consumers need to consider in terms of carbon footprint and impact. Another student created a 3D installation with jam jars and recipes for jam and salsa that her mother makes. Because the student is from Florida, she included a mango salsa as one of the nostalgic tastes of home. Because mangos are not grown in Colorado, her reflection delved into what modifications she could make to the family recipes, for example, swapping in peaches from Colorado's Western Slope.

We and the high school teachers worked together to talk through some of the complexities and learnings and to help students understand the process of critical reflection as their learnings grew. We discussed too how their images and writings would be displayed at the museum show and on social media, and we brainstormed with them other ways the art might circulate as well. The art students created more than seventy pieces of art and accompanying written reflections about their subjects. These were displayed at the Boulder Museum of Contemporary Art; were showcased on The Shed website; and were circulated on students', teachers', parents', and our partners' social media.

Writing students in my Food and Culture course were involved in preparing for the museum show, using their rhetorical skills and awareness of multiple audiences to help me and Constance organize a full-day event at the museum to coincide with Earth Day at the farmers market, which is located just steps from the museum. They created promotional materials and activities for the event, including food literacy documents for the community—lists of farms and CSAs, restaurants that source locally, seasonal gardening information, and other materials. They also created brochures (see figures 5.6a–5.6b) that The Shed used for the rest of its existence as an organization until it disbanded in 2019. Drafts of these materials were vetted by me and other Shed board members, and students worked through multiple drafts as their understandings of audience, voice, and purpose developed.

Students also discussed social media posts and how to best circulate information. The Shed board member Ann Cooper, who has successfully built a powerful social media and website presence through her Chef Ann Foundation, explained in her interview with me for this book, "The best way to get your point across today to people is to be able to do it visually

Our Goal:

To balance our food system through the increased production, consumption, preservation, and appreciation of regional and local food options.

The term "local" isn't just about distance. It is about community. It is about transparency. Being able to ask our local farmers about their crops and how they treat their animals allows us to connect with our food. Knowing that our food is coming from a reliable local source benefits us in so many ways.

The Shed: Boulder County Foodshed is a new coalition of education, non-profit, government, and business leaders in Boulder County.

A foodshed is not just a physical location; it is a **relationship** between people and the environment, between community and its producers, between individuals and the economy.

EATING LOCAL...

- Allows people to reconnect
- Helps save farmland
- Is healthier
- Generates more income for the local economy

Here is a taste of our projects...

Dig In! What is "local" about local food? Is it the value of a connection to others and the earth, humane treatment of animals and workers, money kept in the community, lower carbon footprint, better taste, healthier options, or something else? In order to answer this, we invite you to Dig In! to the concept of local food and what it means to you!

Dig In! is the Shed's countywide educational project to promote conversation about what a foodshed is and how local food relationships matter to our community. The Shed wants to emphasize that not all local food is created equal; just because a product advertises that it is local, does not necessarily mean it carries the same values that we want to promote. Ask questions, dig in to local food!

Visit: theshedbouldercounty.org/dig-in/ for more information and resources to dig in deeper to the conversation

FIGURE 5.6A. A six-panel horizontal folding brochure for The Shed created by author's writing students.

and fairly quickly, and in a way that kind of tugs the heart." Along these lines, students wrote press releases for local and campus news outlets and were interviewed for articles; they wrote letters to potential funders, speakers, and participants to explain the significance of the project. They arranged a "Pollinator Table" with art projects to help teach young children visiting the event about pollinators. They helped organize a full day of

Ecological Community Writing in an Abundant Foodshed

Food affordability is an ongoing problem in Colorado, with 11% of our population living in poverty. The Shed firmly believes that everyone has a right to fresh and nutritious produce, and The Shed is helping to lead the charge for a healthier and more just future.

The goal for this program is to encourage healthy eating lifestyles in our community, support our local farmers, and keep Boulder County's economy local.

DOUBLE UP FOOD BUCKS

The Shed is involved with a program called Double Up Food Bucks. Boulder County SNAP recipients can get twice the value for their dollar on freshly grown Colorado fruits and vegetables at nearby Farmers Markets and local grocers.

DOUBLE YOUR FOOD DOLLARS

Edible Landscapes Project

1) **Community Fruit Rescue** is a coalition project with 350 Boulder County, FallingFruit.org, Boulder Food Rescue and Boulder Bear Coalition. We work with neighborhood coordinators to organize community fruit tree harvests. The **homeowner**, the **harvesters**, and **Boulder Food Rescue recipients** each receive 1/3 of produce gathered.

2) **Food Forests in Parks** works with city planners to integrate edible landscaping in the redesign of park areas and public spaces in Boulder to plant fruit and nut trees and bushes, medicinal plants, and veggie gardens in parks and public areas and provide educational signage and interactive learning opportunities.

3) **Yards to Gardens** collaborates with partner organizations to develop a program to offer consultations for home or business owners interested in transforming their yard into an edible landscape.

FIGURE 5.6B. A six-panel horizontal folding brochure for The Shed created by author's writing students.

speakers, local catering, a seed giveaway from our local seed library, tabling from food nonprofits, and prizes for the art contest winners—all the while adapting their writing to the various audiences to which they presented their information in a variety of genres. Following is one of the local food literacy materials created by my writing students. Through multiple drafts of all of the documents they produced, students worked carefully on tone

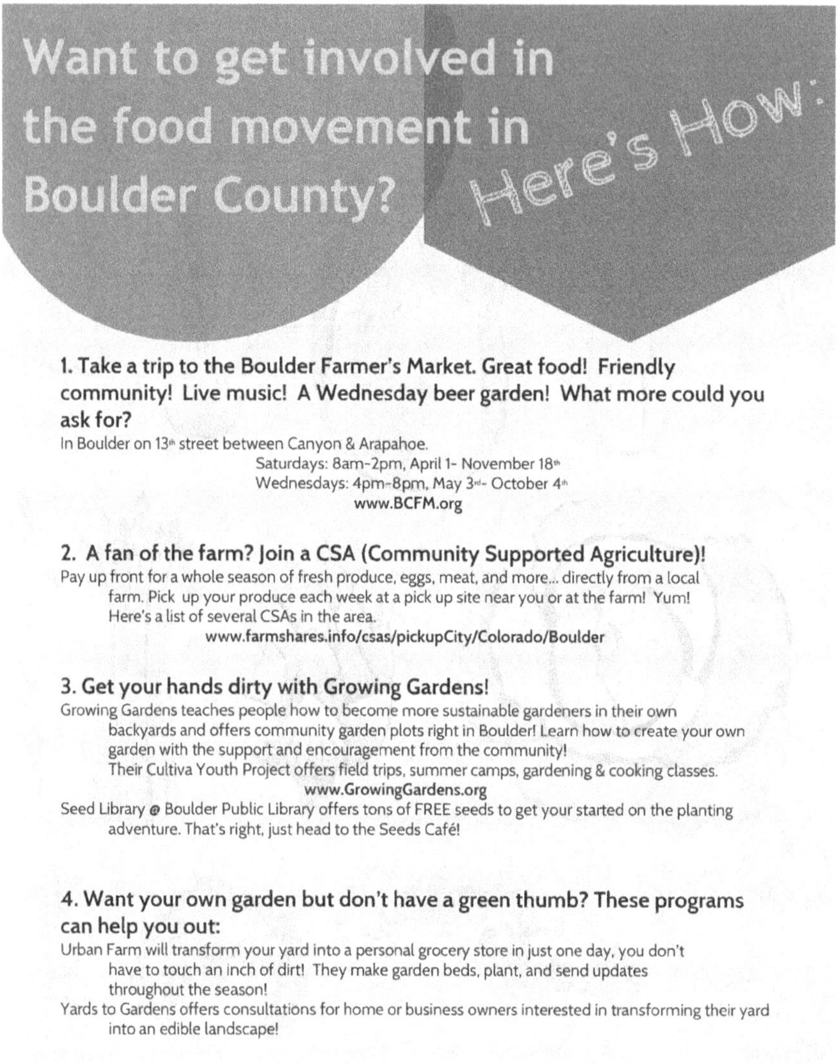

Want to get involved in the food movement in Boulder County? Here's How:

1. Take a trip to the Boulder Farmer's Market. Great food! Friendly community! Live music! A Wednesday beer garden! What more could you ask for?
In Boulder on 13th street between Canyon & Arapahoe.
Saturdays: 8am–2pm, April 1– November 18th
Wednesdays: 4pm–8pm, May 3rd– October 4th
www.BCFM.org

2. A fan of the farm? Join a CSA (Community Supported Agriculture)!
Pay up front for a whole season of fresh produce, eggs, meat, and more... directly from a local farm. Pick up your produce each week at a pick up site near you or at the farm! Yum! Here's a list of several CSAs in the area.
www.farmshares.info/csas/pickupCity/Colorado/Boulder

3. Get your hands dirty with Growing Gardens!
Growing Gardens teaches people how to become more sustainable gardeners in their own backyards and offers community garden plots right in Boulder! Learn how to create your own garden with the support and encouragement from the community!
Their Cultiva Youth Project offers field trips, summer camps, gardening & cooking classes.
www.GrowingGardens.org
Seed Library @ Boulder Public Library offers tons of FREE seeds to get your started on the planting adventure. That's right, just head to the Seeds Café!

4. Want your own garden but don't have a green thumb? These programs can help you out:
Urban Farm will transform your yard into a personal grocery store in just one day, you don't have to touch an inch of dirt! They make garden beds, plant, and send updates throughout the season!
Yards to Gardens offers consultations for home or business owners interested in transforming their yard into an edible landscape!

FIGURE 5.7A. A handout for the art show event gives visitors ideas for how to get involved with local food; content created by author's writing students.

and imagery appropriate for the intended audience. We discussed using rhetorics of abundance versus rhetorics of deficiency or fear. In the "Want to Get Involved . . ." flier, for example, students chose a conversational and upbeat tone with multiple ways to become involved in hopes to inspire fellow community members (see figures 5.7a–5.7b).

5. Double SNAP benefits with Double Up Food Bucks!

The Double Up Food Bucks program doubles the value of federal nutritional (SNAP or food stamps) benefits spent at participating markets and food retail stores, helping people bring home more healthy fruits and vegetables while supporting local farmers. The wins are three-fold: more families have access to fruits and vegetables, local farmers gain new customers, and more food dollars stay in the local economy.

www.DoubleUpColorado.org

6. Do you have yard space but no time to garden? Yards to Gardens has you covered!

Yards to Gardens offers consultations for home or business owners interested in transforming their yard into an edible landscape, based on their interests in having edible veggie gardens, fruit or nut bearing trees, bushes, medicinal plants, bee havens, and drought-tolerant natives.

Contact: micah@350colorado.org

7. Hop on a bike and deliver some good food... before it goes bad!

Boulder Food Rescue focuses on redistributing fresh and healthy foods that would be discarded from local businesses to people of low-income or who are homeless. They also have many projects, such as Garden Against Hunger.

www.BoulderFoodRescue.org

8. Don't feel like getting off the couch? Donate!

Community Food Share works to prevent food insecurity of families and students in poverty by exchanging food from local farmers and grocery stores free of charge. You can also get active and help distribute! www.CommunityFoodShare.org/
Emergency Family Assistance Association provides food and help with critical expenses, like rent, utilities, minor medical costs and transportation.

www.efaa.org

9. It's as easy as 1,2,3,...10! Join the "EAT LOCAL!" Campaign's 10% Pledge!

Take a pledge to spend at least 10% of food money on local foods. So worth it...and so yummy!

www.LocalFoodShift.com/join

10. Netflix and farmer's market popcorn, anyone? Add these documentaries to your queue!

A Place at the Table, Cooked, Cowspiracy, Fed Up, Food Beware, Food Inc, Food Fight, Food Matters, Forks Over Knives, Fresh, Hungry for Change, King Corn, Soul Food Junkies, SuperSize Me, Vegucated... the list goes on!

11. For the restaurant enthusiast, have you tried farm to table?

Here's a list of a few farm to table or sustainably-sourced restaurants in Boulder:

Black Cat	Shine Restaurant & Gathering Place	OAK at Fourteenth
Bramble & Hare	Boulder Dushanbe Teahouse	Leaf Vegetarian
Zeal	The Kitchen	S.A.L.T.
Turley's		

FIGURE 5.7B. A handout for the art show event gives visitors ideas for how to get involved with local food, content created by author's writing students.

The students were enacting *ecological community writing* through these types of writing, through who was and was not contacted for the event, who was and was not invited to speak or participate. Their writing created connections that had not existed before. Their work was ecological. The event was a bustling celebration, visited by several hundred community members

FIGURE 5.8. Visitors look at the high school student artwork. Two- and three-dimensional artwork can be seen as well as plant giveaways and the information table. (Photo credit: Veronica House)

including the high school students, parents, teachers, farmers, local food activists, government employees, and restaurateurs (see figures 5.8 and 5.9).

Several examples of high school student art and accompanying reflections from the museum exhibit follow. These examples from the more than seventy art pieces in the exhibit are representative of the ways in which high school students presented their research through art, their reflections about the research, and the personal connections they were making to much of what they learned. The reflections often illustrate the students' definition building about local food with positive words such as "beauty," "love," "helping," "magical," "abundance," "grateful," and "inspired." The first two examples show students considering connections between family gardens and happiness or health (see figures 5.10–5.11). Other examples show students grappling with difficult ethical questions. For example, one student encourages viewers to consider the cruelty of the poultry industry (see figure 5.12). Several students focused on environmental connections to food. In one of the following examples, after offering facts and numbers

Ecological Community Writing in an Abundant Foodshed

FIGURE 5.9. Ross Rodgers, Founder of Living Seed Library, engages exhibit visitors with seed packets at the information table. In the background, a visitor writes on a white board that asks, "What does Local Food Mean to You?" (Photo credit: Veronica House)

about environmental issues connected to food, the student challenges readers, "How big is your footprint?" (see figure 5.13). In the final two images and reflections, the students depict the nuances of what they saw and felt when visiting a butchery (see figures 5.14–5.15).

When I Eat Raspberries, I Feel Like a Fairy Princess
My inspiration for this piece was the raspberries that grow behind my house in the summer. It's the most local food my family has, a mere 10 feet from the back door. I painted myself as a fairy eating the raspberries because back then, fruit just growing behind our house was magical. I learned how to space everything out in a painting to draw the eye to a single spot in the piece, as well as why it's healthy to the environment to grow my own food. In regards to the foodshed, I believe that everybody with access to ground should have their own garden to reduce the carbon footprint of foods we can grow in our backyards.

FIGURE 5.10. "When I Eat Raspberries, I Feel Like a Fairy Princess" by Bronwyn Ellis is a painting depicting the artist as a child dressed as a fairy eating raspberries from their backyard.

Gardener's Dream: Dad Will Recover
My dad grew up on a ranch near the border of Kansas and Colorado—a wind torn expanse of prairie, mostly occupied by cattle. Every year of my childhood he grew vegetables in our front yard. And every year the garden grew, until the lawn was replaced with beds of tomatoes and chard, and vines climbed up every inch of fence. In January my father was seriously injured in a three-car collision. For two weeks we sat in his hospital room, dreaming about the garden we would plant and my father's voracious tomato plants. Last week he took his first steps without a walker. He walked into the garden and surveyed the barren land, which was beginning to sprout tiny weeds. Inside, in rows of tiny pots, tomato seedlings pushed through the soil.

FIGURE 5.11. "Gardener's Dream: Dad Will Recover" by Jules Weed is a painting that depicts the artist's father in a hospital bed surrounded by family, all imagining the beautiful garden the father would create once he recovered.

"Hen on Straw"
My chicken represents the exploited chicken industry. Chickens are treated without the slightest consideration of their lives, and they are brutally kept and killed before they are sold to big corporations such as MacDonalds and Smithfield, Inc. The importance of eating local is essential, as meat produced more close to home is often more ethically treated

FIGURE 5.12. "Hen on Straw" by Shade Kimball is a sculpture of a white hen placed on a nest of natural straw.

and killed. People fail to think about the maltreated workers and animals, and often processed and unnatural meats lead to diseases such as ecoli and a growing obesity rate.

Miles Matter. Eat Local.
The 12% of the world's population that lives in America and Western Europe account for 60% of the world's private consumption spending while the one third of the world's population that lives in Sub-Saharan Africa and South Asia only account for a mere 3.2%. What makes this staggering difference even scarier is Americans throw away nearly half of the food we buy, adding up to about $165 billion annually, while about 2.8 billion people still struggle to live on around 2 dollars a day. The United States, with only about 5% of the world's population, uses about one fourth of the world's fossil fuel resources. 28% of this energy is used to transport people and goods, such as food. Think about it this way, in 2015 our vehicles drove more than 3 trillion miles. That's enough to travel to the sun and back, 16,000 times. American's effect on the world is astonishing. The miles our food is traveling to get to our plate is causing unimaginable damage. Our footprints are thousands of times bigger than billions of

FIGURE 5.13. "Miles Matter. Eat Local" by Ellia Osofsky is a watercolor painting depicting a large, single footprint creating a black smudge on the planet Earth.

other people around the world, and our food consumption is a key contributing factor. This watercolor piece was done to show our impact on the world in a visual way. It shows a single person's footprint spanning millions of miles, creating one big black smudge on the land we care so deeply about. The piece shows what we are doing to this world with our expensive habits. Our habits are costing us more than just money, they are costing us our home, our world. We are not as doomed as we may seem. By eating locally and consuming only what you need you can reduce your impact and contribute to the helping cause. So I ask you, how big is your footprint?

Raw

My brother works at a local butcher shop called Blackbelly and the farm that provides Blackbelly with most of their meat. I have been to the farm in Longmont that he works on many times and always love looking at all the lambs grazing in the fields, and seldom would think about what happens to all the animals once they leave the farm. When I visited my brother at the Butcher shop, I saw all of the slabs of meat hanging in the freezer room and got a mixed feeling of "ewww!" and "cool!" So naturally I ducked into the back and got as many gross yet interesting photos as I could. I have had conflicting feelings about eating meat for a while, but whether you eat meat or not, I think the most important thing is to acknowledge where it comes from.

This idea is what inspired my drawing, along with the beautiful colors and textures that appear in raw meat if you really look at it. Once I started this piece, I realized how many different layers, shades, textures, and colors were involved in just a tiny section of meat. I have now spent hours looking at and trying to replicate all these aspects, when before I would only encounter animals if they were fully live or fully cooked. When you try to distance yourself from all the steps that happen between the pasture and the dinner table, you lose that connection to the animal and the beauty that they hold even after death. Everything surrounding our culture of meat eating is controversial, with opinions flying around all over the place. There is everything from "Meat is murder" to "Eating animals is the natural way," with every possible variation in between. Doing this drawing has brought up even more questions in my mind of what I think is the "right" food to eat. All I know is, among the flesh and bone in this picture, I find beauty.

Nose to Tail

To shoot for this project, I visited Blackbelly Market in Boulder. Blackbelly is a local restaurant that has their own butcher shop and meat market. The lead butcher, Nate, showed me around and allowed me to photograph his work. He told me all about the work, describing the intricacies of curing salami, locally sourcing meat, and using the whole animal. The title of this

Ecological Community Writing in an Abundant Foodshed   205

FIGURE 5.14. "Raw" by Kate Hefferan is a color drawing of raw hanging meat.

piece came from Blackbelly's motto, "Nose to Tail." When I was visiting, I had the honor of being able to watch them cut up a whole lamb. The lamb was already skinned, drained, and ready to be butchered. I witnessed the whole process from the initial cut to the final clean up. Seeing this unique perspective of meat production was eye opening. In my life, I have exclusively seen meat as perfect slabs of precut protein, sheltered behind glistening sheets of plastic wrap. Seeing the life behind the meat has shifted

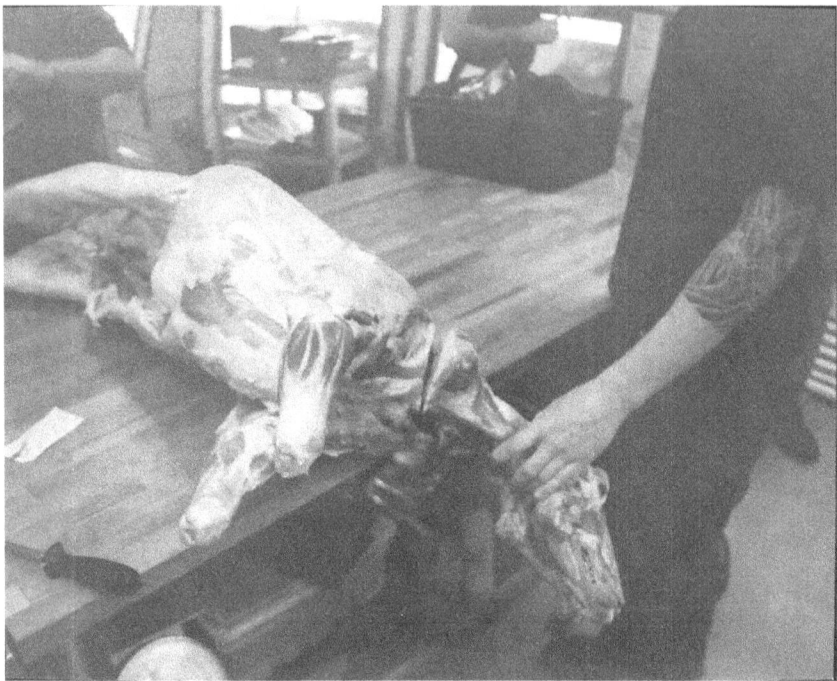

FIGURE 5.15. "Nose to Tail" by Isaac Nagel-Brice is a black-and-white photograph of a lamb about to be butchered.

my perspective and opened my eyes to the beauty and importance of local meat butchery.

The Shed members, exhibit visitors, and guest judges (including a rancher, a teacher, Boulder's mayor, a city council member, and director of food services for the district schools) expressed excitement at the thoughtfulness of the students' art and written reflections, and parents voiced enthusiasm about what their children had learned. In all the events I co-created with various partners across the county, we offered opportunities for writing. Our asset-based approach to *distributed definition building* encouraged an active participation in the creation and sharing of knowledge rather than an outsider or top-down definition. Understanding that food literacy education that offered a narrow set of definitional criteria would also narrow the cultural imagination, we tried to provide open opportunities for people to make the connections between cultural knowledge, desire, availability, relationships, and abundance. Social media shares and rhetorical contagion allow for the circulation of these smaller definitions and stories, as, for example, my students' research and the high school students' art were circulated across social media platforms.

Deepening Definitions of Justice in *Ecological Community Writing*

The joy at the events I've described in the previous pages was palpable, and the beauty of sharing food, seeds, art, and community offered an alternative to industrial, capitalist ways of being in the world. Those emotions were real and vital in exciting event participants to become involved in discussions about local food. In our next class after the art show, my students visibly shared in that community joy, and they were proud of the work they'd done. When we analyzed the high school student art and reflections, as well as the reflective writing community members did at the art show, as rhetorical artifacts, we tried to discern what the learnings were and where further food literacy work could occur.

I asked my students what was present and what was missing from what they thought would have surfaced as part of the definition-building work of the event. They shared ideas such as: Why did some reflections have and others lack justice-focused frames? Why was there more of an individual rather than a community or systemic focus in some of the pieces of student art and reflections? We looked back at the call for art to think through what we might have done differently to elicit reflection that would encourage more of the systemic connections we had studied in our classes.

We were able to consider our role in framing what was and was not available for definition building when we worked with teachers and students. We could use what did and did not show up in the community work happening through these projects. We could challenge ourselves to use what the events revealed in order to consider what kind of a difference the events may have made in peoples' lives and community conversations. We could consider how to better invite people to hold both what we're working against and what we're working toward. We could imagine next steps so that future events better challenge community members to consider whether the conversations we are having reify or disrupt colonial, racist, patriarchal, classist, and ableist paradigms. As bell hooks reminds us, "as democratic educators we have to work to find ways to teach and share knowledge in a manner that does not reinforce existing structures of domination (those of race, gender, class, and religious hierarchies)" (45). This goes for the teaching we do in our communities, as well. There are many layers to *ecological community writing* work in a community. This is where curiosity is so valuable in examination of these events, as my students and I asked what worked well and where we could do better, who and what were and were not present in these projects, who we were and were not accountable to.

My students pointed out that when they discussed local food and created writing opportunities at the Dig In! events, they framed their work in terms of current agricultural production, support of small-scale farmers in the county, regenerative agriculture and environmental issues, community connection, and harms of the industrial food system. All of these issues are incredibly important. My students and I agreed that we did not want to diminish the necessity for these conversations that were the focus of our primary community partner, The Shed, and some of The Shed members' affiliated organizations. But we also wanted to be honest about the possible definitions of local food we may have perpetuated through what we highlighted as key and what we omitted.

We recognized that at these events, we did not discuss some of Boulder County representatives' and residents' conceptions of agricultural heritage as aligning only with white settler families; we did not ask what barriers exist to BIPOC farmers; we did not discuss the history of agriculture in the county related to settler colonialism and Native foodways or current Native food sovereignty work; and we did not discuss, reflect on, and sit in the feelings that arise when we ask what it means to be participating in events about connection to land on unceded land. My students agreed that they would want to offer more focused opportunity for discussion and reflection on these topics with the students, teachers, and community members who participated.

They grappled with the important question of how to deepen the experience and allow people to sit with emotional complexity as those people consider whether a justice-focused element is part of their definition of "local food," and how to do this while keeping people engaged. This is a question without an easy answer, as ecopsychologists have shown. Because *ecological community writing* is iterative and ongoing, we and our students can engage with these challenges, approaching with humility and curiosity the question too of how our collaborative work would need to shift as our understanding of accountability shifts. Deliberately bringing in multiple measures of accountability in our *distributed definition building* method moves us toward that critical space that Mitchell suggests we strive for—when people "become aware of the systemic and institutionalized nature of oppression," performing work with a "goal, ultimately, ... to deconstruct systems of power so the need for service and the inequalities that create and sustain them are dismantled" (Mitchell 50).

My students, partners, and I built opportunities for writing about what "local food" means in order to also distribute the power of who has agency to create definitions, of who has the ability to write. To build coalitions across the local food ecology, we reached out to audiences and stakeholders from schools, farms, restaurants, local food organizations, government, and businesses. We created opportunities for people who might not otherwise meet to come together for a shared purpose at museums, at public schools, at libraries, at parks, and online—to help various publics develop a critical rhetoric around local food in order to support just and accessible local food production, consumption, and appreciation. By tapping into the existing ethos of Boulder County, the environmental concerns, and the conversations around resilience, we invited people to deepen their relationship with our local foodshed, with our producers, with one another, and, to varying degrees of success, with histories included or left out of local rhetorics.

At each event, participants were invited to write responses to questions such as: "Given your experience today, what's one thing you can do to support a more just and accessible local food system in Boulder?" and "What does a strong local food system mean to you? Think about some of what you've learned from the farmers and activists here today." People's willingness to act, and how we encouraged them to "write" themselves into community conversations and action, depended on our framing. We invited people from different kinds of organizations and businesses and neighborhoods, ages, backgrounds, ethnicities, and racial identities to participate so as to visualize that elephant—our own abundant ecology around us. We hoped to offer space for creative and generative ideas to surface that did not allow only for dominant discourses that might drown out others. The thread between the events, the thread helping to build coalitions, was our work to elicit the "writing," broadly defined, to elicit the definition building that connected language, affect, events, organizations, histories, and people together.

The blog posts and written materials, conversations, seed swaps, canning classes, shared information and research, events, frames, all created small stories that we hoped would connect people into a larger narrative unfolding over time. There were so many projects: some might encourage people to think more deeply; some, like the seed saving, harvesting, and canning, were experiential. We were experimenting with lots of different genres and writers and experiences. We did not know what would work or would not work in shifting behavior or attitudes about food. Part of *ecological community writing*'s potential and energy is in this uncertainty and experimental nature. This process of employing *ecological community writing* through *distributed definition building* is how we all begin to see the elephant. It also shows where the blindfolds remain and where we need to be more accountable.

Community gatherings offer space for ongoing transformation of definitions and contrasting viewpoints. The Shed was embedded in an ecology of interorganizational, economic, philosophical, historical, linguistic, and affective entanglements. In Garrett Broad's description of the community workshops Community Services Unlimited organizes in Los Angeles, he explains, "[T]he workshops are never simply about imparting useful information but also about building a critical consciousness to advance a community-led agenda for health and sustainability" (79). He claims, "If local and cultural knowledge systems could be activated . . . and community members were empowered with critical education and skills, personal and community health would become a reality, even in the face of systemic inequity" (80). As climate communicators and ecopsychologists teach us, these moments of activation need to encourage emotional connection, feelings of agency, and community building. In our series of free literacy events spread across the county, people were able to write and speak and create *as* the community (Monberg). Rhetoric and Writing scholars and teachers working with any number of social and environmental issues can directly intervene in community conversations by developing writing and community literacy events and projects for rhetorical saturation, keeping the questions of accountability and a critical, justice-oriented framework in mind.

There needed to be policy change as well, and the fight for systemic change would take time, given multiple and competing visions for local food. I certainly didn't believe that any one event or class would have that kind of immediate impact. But part of our rhetorical work can be to build understanding and capacity within communities and to offer moments for disruption and creation. This involves an "emphasis on unearthing the knowledge and capacities of its participants" (Broad 109). Broad argues that "if community-based initiatives are to effectively advance social justice in the age of neoliberalism," then we must encourage widespread community participation, see the local as part of a global system, and commit to long term activism (175).

In a promising step toward accountability and shifting rhetorical awareness, since I began writing this book, in 2016 Boulder's then–Mayor Suzanne Jones, who was also a Shed board member, signed "A Resolution Declaring the Second Monday of October of Each Year to be Indigenous People's Day." The resolution explicitly details the settler colonial violence and the Boulder County residents' participation in the 1864 Sand Creek Massacre, which the city speaks of in its Staff Land Acknowledgment. The resolution also names important existing Native-led organizations in Boulder. It specifically resolves in Section 6 that "the City of Boulder will work together in partnership with Native Americans to encourage all educational institutions in the city to implement accurate curricula relevant to the traditions, history, and current issues of Indigenous People of and as part of our shared history."

Indigenous People's Day events included food sovereignty webinars and workshops over Zoom during COVID that my students and I attended, led by the First Nations Development Institute in Longmont, Colorado, and Devon Peña's Acequia Institute in San Luis, Colorado. In discussions of destruction of Native food cultures as well as current efforts to preserve Indigenous agricultural knowledge and reintroduce and save heirloom varieties of food and seed, Indigenous land-based literacies are circulated to Indigenous and non-Indigenous participants. Many organizations and farms such as Harvest of All First Nations work every day toward these goals. These discussions are vital in igniting conversations with the City of Boulder to reconceive of how to steward land and counter the ideas circulating that claim that without industrial production, land will go to weeds or will be impossible to cultivate organically or regeneratively on a large scale.

Of course, increased visibility of this work as well as revisions to educational curricula and governmental support, as the city's Staff Land Acknowledgment pledges, must extend far beyond a single day of discussion and acknowledgment each year. The city must ensure that this acknowledgment is what Sarah Ahmed describes as a nonperformative commitment statement (109–14) that moves toward significant, ongoing work to impact how the broader population understands ethical land stewardship, heritage, reparations, Tribal Nation sovereignty, food history on the local level, and what we now call "local food." According to the City of Boulder's government website, there are numerous ongoing "Indigenous Peoples Projects," including "City-Tribal Consultations" with sixteen Tribal Nations, a "Tribal Nation Ethnographic-Education Report . . . informed by in-person interviews with Tribal Representatives," land use considerations, and more ("Indigenous").

In the final years that I was at University of Colorado Boulder, my classes studied an agreement between the City of Boulder and thirteen federally recognized American Indian Tribal Nations, developed and revised collaboratively between 1999 and the present day regarding land use, renaming efforts for public parks and other locations, and reporting to the relevant Native Nations of any objects, burial sites, and other "American Indian Resources" on city Open Space land ("Tribal Nation"). While community members not affiliated with the city government cannot participate in these discussions, I included the published, public information in my course readings so that my students could see where conversations about local food can go, as well as the rich rhetorical life of how people define "local food" unfolding and shifting continuously. Studying this work happening in the community can remind us of the vast knowledges, skills, expertise, linguistic resources, foodways, and relations all around us through an *ecological community writing* methodology that honors multiplicity and complexity.

Since working at the University of Denver, I have been cultivating a relationship with Frontline Farming, a BIPOC- and women-led organization with three urban farms in and around Denver. Their mission "is to create greater equity across our food system on the Front Range of Colorado." "We specifically seek," it states, "to support and create greater leadership and access for women and people of color in our food systems. We achieve these goals through growing food, listening, educating, honoring land and ancestors, policy initiatives, and direct action" (Frontline Farming). In a 2024 collaborative event at the University of Denver, "Writing for Food Justice: Harvesting Hope in Abundance," we worked with my writing classes to educate an audience of DU students and faculty about some of the innovative work happening locally around food access, justice, and sovereignty.

A project partner at Frontline Farming introduced me to several directors and staff in organizations and farms focused specifically on cultural food sovereignty and food justice. I contacted them and others I'd previously worked with: These included Boulder Food Rescue, Harvest of All First Nations, The Ecological Kitchen, and Jefferson County Public Health. Leaders in those organizations participated in a panel discussion. My writing students did all the research and writing involved in helping promote and facilitate the event, funded by the University Writing Program, a Hatherly Family grant, and the Center for Community Engagement to advance Scholarship and Learning.

To prepare for their writing, I shared Dig In! brochures, art, and other writing products from prior projects with my University of Colorado Boulder students, and we discussed how to more concretely center food justice with an emphasis on racial inequities and cultural sovereignty connected to access, hunger, and food waste. Students produced brochures for attendees, researched and introduced our speakers, led the event Q&A after the panel discussion, and created Jeopardy and Kahoot games that encouraged attendees to find answers to questions in their brochures. After the panel, we shared a delicious vegan soul food lunch from local caterer Momma Jah's while panelists sat with students and faculty to continue the conversations in small groups.

Using Mentimeter, in order to get a snapshot of people's understanding by the end of the event, we asked attendees to write words or phrases they find important in a definition of justice-focused local food (see figure 5.16). While some responses such as "supporting local businesses," "environmental," and "connection" resemble those my CU students received while canvassing at the farmers market, framing our event through *critical rhetorics of abundance* encouraged attendees to consider a definition of local food that includes concepts like "equity," "systems change," "liberation," "sovereignty," and "sharing resources." The shift in frames can elicit shifts in definition.

The "Writing for Food Justice: Harvesting Hope in Abundance," event helps to illustrate that when we consider creating spaces for *distributed definition building*, careful framing on our part is crucial. The food literacy elements that we include or exclude in each project and event can help attendees delve into complexities as they consider the question of how to define "local food." In using this method with many complex issues, we are

FIGURE 5.16. Word Cloud Generated using the Mentimeter app is based on fifty-three attendees' responses to the question of how they would define justice-focused local food.

not imposing a single definition, but we are not supporting an anything-goes concept of *distributed definition building* either. We are guiding people into complexity to support educated, thoughtful engagement.

Dynamic Definitions in Dynamic Ecologies

When community writing practitioners encourage writing in communities, our own and our students' work is only part of the ecology. It is extremely rare (and Paul Feigenbaum points out how important it is to express this to students) that one person or piece of writing will significantly change a system (*Collaborative* 109–11). It is important that students understand their writing and projects as happening within interlocking systems. Students can understand their writing and their work as ecological—it will connect or link with other people and writings in ways we perhaps cannot imagine. Direct impact is a tricky thing to assess. It is hard to know what happens when someone picks up a brochure at an event, or when they write on one of the large easels at our events a response to a reflection question. Who knows how a high school student art piece or a carefully framed message on a bumper sticker might make a viewer think anew, or how a city's land acknowledgment might tip someone into reflective action.

Understanding that textual production doesn't automatically equate to circulation, impact, or relationships, The Shed board and the affiliated organizations, offices, and partners were encouraged to circulate the writing products and art images widely and to encourage their followers to circulate, and so on. Participants at events were encouraged to bring their developing literacies back to their communities, and we hoped that no matter what definition of local food people created for themselves, it would be thoughtful and that, perhaps most important, people would feel agency to write and continue to deepen that definition for themselves while acknowledging the abundant ecology within which they do so.

As I've discussed in this chapter, the work coming out of our events and projects can make us as community writing practitioners think anew as well, can challenge us to rethink our practices, to shift our focus for future projects, and to continually consider those we are accountable to. This work is messy, and it is never done.

University writing students helped to co-write the community's definition for "local food" through their data collection and presentations to various audiences around the county. But their writing is not privileged; it is part of and helps build a dynamic ecology of writing. The high school students co-wrote the definition through their art projects and reflections; the farmers co-wrote through their survey responses and oral histories; the farmers market patrons co-wrote through their responses on index cards. Boulder County itself has an ethos, built on a history of environmental activism to protect City and County Open Space land—those historical commitments also co-write the community's definition and vision for local food. The county and city are also built on a history of violence against Native peoples, and that history co-writes what current rhetorics of local food make possible, as well as what changed, anticolonial rhetorics would open as possibilities for how we feed ourselves and our communities. And as I will suggest in the next chapter, the land continuously co-writes, the climate co-writes, our water rights laws co-write, the seeds and plants and animals co-write, because these other-than-humans cannot be ignored in any meaningful discussion of how local food can thrive in Boulder County.

Future students and involved community members will continue to write and revise local food's definition in Boulder County. To ignore this complex, evolving ecology in our structuring of community-based projects and events would be detrimental not only for our students' learning about the nature of writing's function in the world but also for the many, many people implicated in the ecology of partnerships impacted by the range of projects. Seeing any one of the elements in isolation is to ignore the dynamic ecology. We are reminded by adrienne maree brown:

> At a certain point, even if collective action feels far away, there has to be an awareness of the pattern. We have to develop the systemic intuition to sense that the same glitch is present throughout all the systems. Thinking that your choices only impact you or those you immediately know—that you needn't be concerned with or accountable for the results—is supremacist thinking at the root. It gets packaged as freedom and independence, but we are not individual entities. Humans, like all of nature, live within systems of relationship and resource. Our freedom is relational. ("emergent")

Helping students understand and practice how writing and rhetoric can be used as a part of community building for collective action in dynamic systems of relationships and resources can be an important goal for Rhetoric and Writing courses.

The Shed, project partners, students, and I brought together different constituencies to look at collective impact and develop a common agenda, which is not a shared definition but rather shared literate, affective commitment and active involvement on the part of our community. Through all our collaborative work across events, materials, and groups, we hoped to saturate our community with a *feeling* of and opportunities for engagement through an abundance of food, resources, knowledges, skills, relationships, concrete steps of further action, connection, and complex emotions as we consider why local food systems matter.

In the process of facilitating this distributed, ecological approach to understanding local food's meaning in Boulder County, the conversation shifted, including among some of The Shed board members. We were not talking about "local food" with a fixed definition anymore, and we weren't talking about it only in terms of distance. Boulder's historical battles around land, housing costs, laws and policies, people, language, intersecting institutions and organizations, movements in food sovereignty and regenerative agriculture, the climate, water, and myriad other factors all influence the ways in which "local food"—and the multiple forms of work to promote its production, sale, accessibility, and popularity—is defined in Boulder County.

This distributed definition of "local food" is written through my students' guest presentations to food-related classes across our campus, through their research and multimodal materials that were shared with community partners, and through their co-creation of events to support public literacy. It is written through my community project partners' work in their nonprofits, on their farms, in work on policy to "cultivat[e] cultures of collaboration" and to provide people with information about how to participate (Levkoe et al. 56). For farmers, it is written through their embodied and text-based literacies of soil, plants, water, and farming practices—all essential local knowledge to be shared, preserved, and circulated. It is written by Native Nations members, whose ancestors' foodways were destroyed and land was stolen, but who nevertheless are working to preserve and revive ancestral ways of knowing and being with land. It is written by community members who in so many ways across time and space collaboratively build, support, and redefine local knowledges and communities, writing ideas of "local food," exposing and highlighting what is good and what must change. It is written through the rhetorical life of the definition of "local food," again and again, transforming, building, informing, transforming again endlessly, with writing connecting us all.

6

Abundant Partnership

A LOCAL, ORGANIC APPROACH TO RHETORIC AND WRITING STUDIES[1]

with Kerry Francis, Tim Francis, and Kelly Zepelin

> *To be native to a place we must learn to speak its language.*
> —ROBIN WALL KIMMERER, *BRAIDING SWEETGRASS*

> *There is a relationship of the body to the land and to the histories connecting us all.*
> —JOYCE RAIN ANDERSON, "REMAPPING SETTLER COLONIAL TERRITORIES: BRINGING LOCAL NATIVE KNOWLEDGE INTO THE CLASSROOM"

> *Where does your own self end and where does the air or water begin? Is the air inside my lungs me? The oxygen in my blood? Is the air I exhale not me? Is the water that I drink, that comes into my cells not me?*
> —PER ESPEN STOKNES, *WHAT WE THINK ABOUT WHEN WE TRY NOT TO THINK ABOUT GLOBAL WARMING*

Building a just and resilient local food system in the face of extractive corporate production methods and systemic inequities in who has access to

---

1. I am deeply thankful for Kelly Zepelin, a wild food forager and cultural anthropologist living in Durango, Colorado, without whom this chapter would look very different. Kelly and I collaboratively wrote an early version of this for our chapter, "When the Land Writes: The Rhetorical and Reciprocal Lives of Soil and Plants" in the edited collection *Food Justice Activism and Pedagogies: Literacies and Rhetorics for Transforming Food Systems in Local and Transnational Contexts*, edited by Eileen E. Schell, Dianna Winslow, and Pritisha Shrestha. My collaboration with Kelly and her generosity in teaching me about foraging in Colorado is another manifestation of abundance. I am also grateful to Eileen, Dianna, and Pritisha for their thoughtful comments that helped shape our thinking as we revised that chapter. Tim Francis and Kerry Francis read through and edited multiple versions of our initial interview and this chapter. I shared the writing about Dharma's Garden with them as a collaborative process that has shaped my thinking in beautiful ways. Finally, I am thankful for Iris Ruiz's feedback on this chapter and for her scholarship that has impacted my thinking and practice.

https://doi.org/10.7330/9781646427208.c006

healthy, culturally sustaining foods requires creative thinking from a broad ecology of actors and an expansion of how we build partnerships and connections through our teaching, advocacy, and research. An *ecological community writing* approach means supporting production of texts and food literacy projects that allow for interdisciplinary and trans-community partnerships and engagement. It means encouraging the circulation of writing and ideas emerging from people of different ages, abilities, backgrounds, cultures, income levels, expertise, and races, some of whom may never have thought of themselves as writers. And, with as capacious a view of an ecology as possible, it can mean expanding our definition of writing beyond a humans-only activity.

As I close this book on *ecological community writing*, I broaden my sense of community even farther, offering another example of collaborative writing that has grown out of an ecological approach to knowledge production. I turn to my work with two farmers in Boulder as I look into a transformational practice that depends on reciprocal collaboration with other-than-humans. In so doing, I urge us to consider how the method of *distributed definition building* and the concept of partnership grow if we ask how other-than-humans write.

Community literacy and food literacy scholarship center reciprocity. Indeed, the fall 2019 special issue of *Community Literacy Journal* on food and environmental justice takes reciprocity as its central tenet (Opel and Sackey). When community-engaged writing studies teachers and scholars discuss reciprocity, we typically refer to human partnerships between community members, nonprofit staff, faculty, and students (e.g., Cushman, "Rhetorician"; Del Hierro et al.; Parks; K. Powell and Takayoshi; Shah). I encourage an expansion of our community partnership model and believe that when we consider other-than-humans as our partners in definition building, we open ourselves to their rhetorical and authorial agency and knowledge.

Indigenous scholars have long focused on the agency of other-than-human beings as well as on the importance of relationship and reciprocity (e.g., M. Powell 41; Riley Mukavetz and Powell, "Becoming Relations"). It is critical that non-Indigenous community literacy scholars and teachers like myself acknowledge what our Indigenous colleagues have been reminding us for decades: that these foundational principles are not new but come from long-practiced and long-theorized Indigenous knowledges, or what Vine Deloria Jr. explains as Indian metaphysics, "the realization that the world, and all its possible experiences, constituted a social reality, a fabric of life in which everything had the possibility of intimate knowing relationships because, ultimately, everything was related" (in Deloria and Wildcat 2). I want to, as Lisa King, Rose Gubele, and Joyce Rain Anderson have urged, "recognize and honor the intellectual work of indigenous thinkers and rhetoricians who have carried this knowledge into the present ... Ultimately, this work continues the call for alliance among Native and non-Native scholars, teachers, and students to transform the discipline for the benefit of all" (14–15). In this spirit of recognition and striving toward alliance, I offer a theory of partnership in community-engaged projects and a shift in food literacy studies to include other-than-human elements such as plants and land as partners that write *about*, *for*, and *with* us, to play on Thomas Deans's often-cited "three paradigms for community writing" (*Writing Partnerships*).

As our conception of an ecology continues to widen from classroom to single partnerships to multiple partnerships spanning disciplines, campuses, and communities, our conception of reciprocal partnership can deepen as well. When reciprocal partnership is considered through an anticolonial lens, it centers Indigenous teachings of other-than-human knowledges often marginalized or delegitimized in white-dominant institutions like industrial food and seed companies and academia. In an expansion of both our field's understanding of reciprocal community partnership and my methodology of *ecological community writing*, I am guided by the many Indigenous-led decolonization and re-indigenization efforts underway both within and outside of Rhetoric and Writing Studies that are revealing and remembering knowledges long suppressed by settler colonial paradigms, including efforts by Andrea Riley Mukavetz and Cindy Tekobbe, Gabriel Ríos, Joyce Rain Anderson, Sean Sherman, Devon Peña, and many more.

In highlighting some of the work happening with farmers in Boulder County, I share approaches to local food work that incorporate other-than-humans as writers that help us define what is just, appropriate, and possible within our ecology and foodshed. I offer from-the-ground examples of this expanded definition of writing to include nonalphabetic, nontextual, and beyond-human creative agency. Through these examples, I posit other-than-human agential creation as a re-conception of writing and authorship—one that I see as aligned with a larger Indigenous-led movement in decolonizing academia and Western ways of knowing, such as Iris Ruiz and Sonia Arellano's conception of "quilting as method," Kimmerer's discussion of the limits of Western science (*Braiding*), Linda Tuhiwai Smith's work on "decolonizing methodologies," and Qwo-Li Driskill's concept of "decolonial skillshare." As Lisa King reminds us, "the people, the culture, and the land take their meanings from each other" (19). Like King, Vanessa Watts reflects on the relationship between the agency of land and humans. In "Indigenous Place-Thought and Agency Among Humans and Non-Humans," she explains the Anishinaabe understanding of human agency as coming from our relationship to the personhood of the land. "Place-Thought," she writes, "is based upon the premise that land is alive and thinking and that humans and non-humans derive agency through the extensions of these thoughts" (21). People, cultures, land, and all beings are co-creating always together.

On the other hand, Western philosophical practice "enables the naturalization of the treatment of land and other non-human beings as resources, to be mapped, known, and cataloged as objects, and owned by human knowing subjects" (Rosiek et al. 339). This vision of land and other beings as resources to be owned is precisely the hoarding principle I've discussed throughout the book as cultural sickness in the United States, a definition of abundance framed through individualism; accumulation of resources, wealth, and power; and an exploitation of other humans, animals, and the earth. This ideology is also grounded in how we define things. Time and again, those who get to make the definitions that frame our world wield the power to reenforce or reject cultural paradigms and ideologies. By decentering Western methodologies and definitions of writing, we can recognize that other-than-humans partner in creation and help to shape our actions.

Humans have long struggled to find language to describe what this partnership is and to explain what it is that the other-than-human world does in its acts of creation. Scholars have used terms such as "agency," "language," "rhetorical life," and "grammar" to describe the creative work of other-than-humans. When I use "writing" and "authorship" in relation to what other-than-humans do, I hope that readers understand that I too am grappling with language inherited by Western, colonial epistemologies. I could have chosen, for example, cultivating or growing, to draw on more natural terms, or co-creating or communicating or co-shaping. The English language has no word for the creation I seek to explore. As Kimmerer says,

"English doesn't give us many tools for incorporating respect for animacy" (*Braiding* 56). So, I choose "writing," as flawed or partial as the word may be (and welcome people from other fields or positionalities to choose a different word to get at this inarticulable concept), in order to urge the field of Rhetoric and Writing Studies to continue to re-vise, as in "to see anew," writing's potential, its scope, its very essence. Deloria asks his readers to remember the "grievous sin of the Western mind: misplaced concreteness—the desire to absolutize what are but tenuous conclusions. . . . [I]n this reduction of knowledge of phenomena to a sterile, abstract concept, much is lost that cannot be retrieved" (in Deloria and Wildcat 6). Through an ecological lens, I envision writing as an animate, always-in-creation thread that comes from and connects things, a capacious view that can help Rhetoric and Writing faculty and students understand partnership work with other-than-humans in beautifully rich and unexpected ways.

In expanding the community writing partnership model to include the authorial presence of the other-than-human, we can attune to community building and justice-oriented work literally from the ground up, from the grassroots. This expansion is part of a larger project that expands food justice efforts to focus more explicitly on racial and decolonial justice and the Black, Latinx, immigrant, and Indigenous knowledges essential to building just local food systems (Garth and Reese 1–28). As García and Baca posit, "epistemic alternatives . . . can move us beyond Western categories of epistemology, thought, and feeling" as part of the delinking project they encourage (2). As I consider my own responsibilities to challenge the field's "story-so-far" of what writing is, I ask how I can embrace García and Baca's call for "pushing the 'discipline' to move beyond itself," from what Ruiz and Baca have called the field's "colonial unconscious" (García and Baca 24; Ruiz and Baca 226). In *Braiding Sweetgrass*, Kimmerer describes a "grammar of animacy," in which, when "listening in wild spaces, we are audience to conversations in a language not our own" (48). She calls us to "imagine the access we would have to different perspectives, the things we might see through other eyes, the wisdom that surrounds us. . . . [T]here are intelligences other than our own, teachers all around us" (58).

For decades, community-engaged scholarship has urged academics to recognize assets, knowledges, and expertise beyond the confines of our campuses; to value, cite, and share knowledges within and across communities of humans. As I've suggested in the pages of this book, I know that I function within a vast ecology of people working and organizing and writing for change. An assets-based approach shifts us, as Rachael W. Shah explains, to recognizing "the gifts of individuals, including especially populations traditionally framed in terms of their deficits" (23). An *ecological community writing* methodology actively builds coalitions and connections through distributed writing projects that recognize people's assets.

What if, as Kimmerer's grammar of animacy suggests, we approach land, seeds, animals, and plants in a similar way, valuing and sharing "grammars" across species and things to include other-than-humans as rhetorically active, knowledge producing collaborators that can participate in *distributed definition building* of what local food can be? What knowledge do they already have, what are they ready to share as gifts to us (Kimmerer, "Serviceberry"), and what do they want of us?

When I acknowledge other-than-human rhetorical life and authorial agency, I am influenced by Gabriela Ríos's theory of land-based rhetorics: "[L]and-based literacies are literal acts of interpretation and communication that grow out of active participation with land.... Indigenous relationality recognizes that humans and the environment are in a relationship that is co-constituted and not just interdependent. Additionally, Indigenous relationality recognizes the environment's capacity to produce relation" (64). She explains that this "relationship between land and bodies ... produces knowledge" that binds us and the land—inspiring a cycle of reciprocity (65). Similarly, Angayuqaq Oscar Kawagley explains that his Yupiaq language "contains the creatures, plants and elements of nature that have named and defined themselves to my ancestors and are naming and defining themselves to me. My ancestors made my language from nature" (1). Given these concepts of relationality, of rhetoric's force to catalyze change, and of the ways in which language derives from nature, we can attune to knowledges produced between humans and other-than-humans when working toward a just local food system. We can consider how our relationship with a plant produces knowledge or action, how the land's response to our activities on it *writes about* the benefits or detriments of our actions. We can consider how things co-create relationships and languages and definitions that change our actions and that elicit consequences.

I imagine that some readers may balk at the idea of other-than-humans as writers because it seems like anthropomorphizing. I am once again guided by Kimmerer to suggest that perhaps that is simply a Westernized view of what I'm suggesting. In the interview "We've Forgotten How to Listen to Plants," Kimmerer explains her use of "personhood" in discussions of other-than-humans. She says,

> When I say that this aspen tree is a person, I mean it. . . . [Y]es, this is a sentient being, a conscious being, a being with its own intelligence. You know, Western science tells us, "Oh, you must not anthropomorphize." I'm not anthropomorphizing, I'm botanizing. That aspen is its own kind of person. It's not like they're human people. They're aspen people.

Learning to read and circulate and "co-constitute," as Gabriela Ríos calls it (64), these various forms of writing (not human writing, but land writing) is part of our work as community-engaged scholars, teachers, and students. I acknowledge Kimmerer's concept of the grammar of animacy and Kawagley's explanation of elements of nature naming and defining themselves. Broadening the scope of *distributed definition building* to recognize authorship in other-than-human beings begins to shed Western and settler colonial conceptions of hierarchies of knowledge and domination and instead centers Indigenous paradigms, scholars, and activists who can in turn inspire collective movement toward a more just food system and a more just discipline.

Distributed Definition Building: Cultivating Reciprocity

I am sitting at a picnic table in one of my favorite spots in Boulder, the small neighborhood farm called Dharma's Garden that I mentioned in the previous chapter. Founding farmers Tim Francis and Kerry Francis, who are caucasian from European ancestry, turned just one-half acre of this five-acre lot into a productive market garden and have helped feed my family for years through a farm membership that includes weekly vegetables during harvest season. I've been coming here for a decade, first with my young daughters who walked the meadows and participated in seasonal festivals as preschoolers from the school next door. Kerry was my younger daughter's beloved preschool teacher. As I got to know Kerry and Tim better, I would bring my university writing students to experience the land's abundance and to learn about ethical land stewardship and urban regenerative farming. As our partnership deepened, I would help purchase supplies for community education workshops at the farm on food preservation and heritage seed saving through the Dig In! grant from University of Colorado Boulder's Office of Outreach and Engagement. Tim and Kerry received a grant from The Shed to purchase a new pump system for their duck pond that would be used to irrigate and fertilize the farm. Often on neighborhood walks with my dogs, I'll see Tim up in a tree tending a nest he built for resident owls; or Kerry teaching children, bent over flowers or running along the paths that wind through the land; or a group of volunteers out in the garden, working the rows, while a herd of sheep cut and fertilize the grasses as they eat in the meadows.

In a neighborhood quickly gentrifying with multimillion-dollar homes pushing out longtime residents, and with high property taxes and cost of living threatening the large mobile home and affordable housing communities a few streets north of the farm, Dharma's Garden is an oasis in the city. In 2021 after a multiyear campaign to save the farm from encroaching development of more multimillion-dollar homes, over 600 individuals and families in the community donated to help Tim and Kerry's nonprofit successfully purchase the land they had leased for years to be held in perpetuity without risk of development. It was an example of community coming together and preserving the land against seemingly insurmountable odds.

Rather than feeding into "cultural systems that produce both environmental degradation and racial and economic inequality," Kerry and Tim have moved away from a common CSA model in order to even more deeply nurture relationship and reciprocity (Alkon and Agyeman 9). People pay what they can up front in the winter to come to the farm each week during the harvest season and take as much as they need for the week from what is available. Some community members give $5 for the season, others upwards of $1,000. Kerry and Tim's vision is to be "a project OF the community, BY the community, and FOR the community" (website; original caps). As they explain on the Dharma's Garden website, "[w]e're doing things a little differently here. We are not about producing an agricultural commodity to sell; rather, we are focused on sharing an *experience of tending the land* with our community. In the spirit of true sharing, at our on-site weekly market, all Dharma's Garden MEMBERS are welcome to take home as many of the fruits and vegetables as will fulfill their needs. We ask only that you use what you take home with you, so that nothing goes to waste. The model is simple: **TAKE what you NEED, and GIVE what you CAN**. For this model to work, it is important that those of us who *can* afford to

FIGURE 6.1. Kerry Francis on the far left smiles at the camera along with farm volunteers. They are harvesting flowers and vegetables at Dharma's Garden. (Photo Credit: Dharma's Garden)

do so, *give generously*. In order to cover the significant costs associated with tending the land to produce this shared abundance, we rely on the generous contributions from our community" (bold, caps, and italics in the original).

Tim and Kerry love this land, and it seems to love them back, giving of itself to them as they give themselves to it. Theirs is a model for feeding community and connecting to land not built on extraction or domination or exclusion based on income level. I view them and the land as collaboratively composing what is possible for feeding our neighbors and nurturing connection to the earth (see figure 6.1). In the beauty of this coauthorship, this *distributed definition building*, are possibilities for expanding our understanding of deep reciprocity.

Here on the farm on a comfortable July day, I talk with Kerry and Tim about building partnership with land. I begin by asking what drew them here. Tim explains, "There was this calling, this longing that started quietly, and we didn't know what to make of it at first. We just felt drawn in toward *this* land without knowing what it meant or what would come next. It's very hard to put into words because it's so beyond words, it's pre-words." They've told me and my visiting students before that the land has an agency that expresses itself through their work with it. Kerry says, "Once we lived *with* the land and lived *on* the land and worked the soil every season, that then led to a kind of conversation. That's also really hard to put into words, but it was some kind of communication where we felt like we could sense something of what the land wanted or that we were being guided in some way. It is a personal relationship and conversation with the land that took time. And then from that conversation arose a vision that got clearer and clearer."

Tim jumps in, "Yes, and I think what we began to experience was an inner receptiveness to a vision of the land that was not solely our own. There was a very real sense that the vision itself arose out of the land. And I think this is partly what makes it hard to put into words, because there's a danger of anthropomorphizing, as if the land is a humanlike person that has its vision and its wants and desires. But it's much deeper than that. It feels to me so relational."

"Not like the land has the vision, per se," Kerry tries to explain. "Rather, we listen, and we become part of a co-created relationship." Tim adds, "We—the humans tending the land—are in this living conversation with the land that is supporting us; and through that the true vision reveals itself."

This communication that they describe is hard to define, harkening back to Kimmerer's awareness of how difficult it is in English to find language for other-than-human animacy—it isn't spoken or textual, of course. There is a rhetorical vibrancy that comes through what the land grows and doesn't grow, through which animals thrive on it and where. I'm reminded of Kimmerer's observation on reciprocity, "Something essential happens in a vegetable garden. It's a place where if you can't say 'I love you' out loud, you can say it in seeds. And the land will reciprocate, in beans" (*Braiding* 127).

My former writing student Madeline Nall, whom I quoted in the book's introduction, first came to Dharma's Garden on a field trip with our writing class in 2017 and served as community engagement coordinator for the farm for the next five years. In our interview for this book about their experience with the land, they explain, "There were explosions of flowers and veggies everywhere, existing harmoniously with the natural dynamics of the land. You can see and you can feel how happy the land is to grow that food. The interconnectedness between human and earth's vitality was potent. And the way that I'm speaking about this too, is totally a reflection of how much the land has impacted me, because I would never have been personifying land before this experience. And I'm like, 'Oh, the land wants to do this.'" I ask Madeline what they mean, and their response echoes Kimmerer's notion of unspoken messages of love that flow between beings. Madeline explains, "The produce is huge and vibrant and vivid, and the carrots are just bursting with flavor. There's something so surreal about picking a carrot and then eating it right out of the ground. It's as fresh as it can get, and it's totally portrayed in the flavor palette. For many visitors like myself, the act of eating and tasting the food grown here is a way to access that base feeling of connection to nature: the symbiotic relationship of tending to the garden and the garden nourishing our bodies. But a little less tangible is the animal life. I was just there last night and, you know, there are the two owls. They fledged, but they're still hanging out, and they were just sitting on the fence post with the garden. And then we had a mother deer with her two babies coming by, and tons of bird life. It's clear that the land serves as a haven for humans and animals alike, and I think that's why it's like, the relationship with the land is so much more than what we can harvest from it. The land and the garden remind us that we are part of this complex, thriving ecosystem, and each being serves a critical role."

Tim, who has been a mentor to Madeline, feels that a disconnection from the natural world that sustains us all is one of the biggest problems so many people face and one that perpetuates legacies of extraction and injustice. My re-conception of writing partnerships aligns with Ruiz and Arellano's vision for "medicinal rhetorics" in that it "heals that which was severed through colonial relations and colonial renditions of the human relationship to the Earth" (152). In other words, calling this land-based or plant-based agency "writing," as inadequate as the word may be, is an attempt at healing the rift by acknowledging the co-creative partnership between us that is so often forgotten and that is a direct outcome of our separation from the land that leads to so much exploitation and destruction.

Tim explains, "If we as a human society were really to honestly have that kind of conversation or relationship with land, we would find that sometimes a particular piece of land is well served by a garden, or maybe it needs to just stay like it is and not have even a garden on it, maybe not have humans on it at all—just remain in a wild state for other creatures to make their homes or pass through. More often than not, we make our decisions about land use based on some utilitarian calculations, ignoring any kind of real relationship with the land. I would argue that's the very reason that we're so deeply entrenched into so many profoundly disturbing crises around the world today—from climate destabilization to pollution to habitat destruction, and everything else.

"When we first had this calling here, we didn't quite know what would develop. And then this idea of having a garden just seemed natural, but where would it be? It wasn't until we were here for a little while that we were drawn to where the borders of the garden would be. And there was certainly some logical reasoning behind it, but it's beyond just logical reasoning. It came out of conversation with the land. Again, it's hard to express, hard to articulate—it seems prior to words and human language. It's more of an intuitive feeling. The creative process of developing our garden started with the felt sense of the garden, then the vision of the garden began to coalesce around that. And as the human will began to engage, to pursue that vision, it looped right back in again to the felt sense: 'does this feel contrary to what's happening already, or does it seamlessly merge with what's already here?'"

The land *writes for and with* us in partnership. When we expand *ecological community writing* to include other-than-humans, they participate in *distributed definition building* by responding to our questions and actions in ways that can alter or confirm what we do.

"People ask us if we want to build out the farm more. That's really interesting because—and this is a good example of why it's so important, *so important*, to have this relationship with the land, and to actually live on the land," Tim explains, "because it was only through being here and listening to the land that we came to understand the real sacredness of the wild places and the value of finding that balance between the wild and the cultivated. Probably a more typical way to go about starting a new community farm would be to first start with the idea, and then to go find a piece of land to impose that idea upon. Maybe you pull up a Google satellite view of the property and start planning, 'here is where we could put this or do that.' That's the standard way that our modern society works with land. More often than not, I think you'd find that the whole project would start to go astray, because without the foundation of being with the land and listening and building relationship, all we have are our human intellectual ideas of how to use the land to satisfy our own goals. Even if we happen to have the altruistic goal of preserving wild spaces or promoting regenerative agriculture, our whole relationship with the land would be extractive in nature—we would be extracting from the land its resources to meet the needs of this *idea* that we have. And I think this is where human civilization as a whole has gone astray.

"What we discovered here," Tim continues, "is that there is another way of being with land, working with land, that turns that paradigm on its head. And to be clear, this isn't new in any way. This is an ancient way of just being with land. With any piece of land that you're trying to work, you leave a part of it wild. And then you can observe what's happening there without any human inputs. And you let that help inform the rest of the project. That also is not a new idea, but more a forgotten one.

"And actually, even the idea of *untouched wild places* is built upon a premise that is fundamentally wrong," Tim explains. "National parks are an example of wild spaces that we have preserved, and I think it's important to realize that this common understanding of national parks is a lie. Many—if not most—of the places that we have now made into national parks were inhabited by Indigenous peoples, and too often, those people were forcibly removed to make way for the newly designated parks. Those displaced peoples had lived in those places in *relationship* with the land, and in *relationship* with all of the other beings who also lived there. The very idea that there are two types of places—those that are wild (devoid of human influence) and those that are cultivated (shaped by human influence)—is indicative of a profound error in our modern way of thinking about the world, and we see it throughout our modern society. It is this mistaken idea of human beings as separate from the natural world."

Tim's idea echoes many Indigenous scholars who have written about the concept of "wild" as one that doesn't exist in most Native cultures. Mistinguette Smith says that the concept of wild land "requires a colonial memory, one unwilling to witness long histories of indigenous naming, habitation, settlement, and cultivation" (138). Enrique Salmón has written extensively on this topic: "There is no word for the concept of *wild* in [his] native language of Rarámuri," and in fact, he says that he has found no Native language that includes a word for the concept of *wild* ("No Word" 24, 27), elaborating: "A Rarámuri worldview does not differentiate or separate ontological spaces beyond and between the human and nonhuman worlds. We feel that we are directly related to everything around us. The trees are us; we are the trees. I am rain; rain is me" (25). He posits an "interconnected view of all life as a kincentric ecology, in which everything is a relative" (25–26).

Tim and Kerry's ideas of relationality and for rhetorical listening and building relationship with the land before imposing their own ideas on it also align with scholarship on best practices in partnership building and community listening (Cushman, "Rhetorician"; Goldblatt; Mathieu; Fishman and Rosenberg). Like Paul Feigenbaum's conception of collaboratively imagining with our human partners (*Collaborative* 44), Tim and Kerry collaboratively imagine with the land (and animals and plants) what might be possible for the land and what is being asked of them by the land. This notion also connects to the idea of radical imagination, which is meant to be a collaborative act. Here, we can see that radically imagining and co-creating a more just future can involve collaboration with other-than-humans. It is not through literal talking or writing back and forth, but, as Tim mentions, there is an exchange happening that is beyond human language.

There is an other-than-human rhetorical argument being made for the various choices Kerry and Tim then enact. Tim points to a patch of grass next to where we are sitting: "Even right here, this little patch of tall grass. For years, we had been mowing it. And then we just realized that it was a different kind of grass from all the other grasses around. And whenever we'd mow it, it would look all hacked-up, as if it was resisting being mowed. And it seemed like maybe it was trying to grow a lot taller than all the other grasses around it. So, finally, we thought, okay, let's just let it grow taller. And now, there's actually a lot of life that happens there. There are all kinds of insects that live in that little patch of grass. We see dragonflies there, and we see birds going in and getting seeds. It's no hassle to us to have one little patch of tall grass in the middle of our mowed area. And it's also nice to have some areas that *are* mowed, so that we can sit here at this table like we are now and have a conversation. So, we just have these decisions that are made out of observation and relationship."

Anthropologist and farmer Devon Peña might refer to this as an example of "relational solidarity with self-willing land+water as vibrant co-actants and shapers of the world" (Peña 93). The more time I spend with Kerry and Tim, the clearer it becomes that there is rhetorical agency in this grass, the foxes and owls, and the vibrant produce. They are creating in solidarity along with Tim and Kerry, as co–world shapers. They are involved in non-alphabetic, agential co-authorship of what is possible on this land, and in this way, the reciprocal relationship, which "extends temporally," "means doing the work of revisiting structures of power in continual conversation"

(Bernardo and Monberg 85). bell hooks reminds us that as an antiracist practice, "[t]o build community requires vigilant awareness of the work we must continually do to undermine all the socialization that leads us to behave in ways that perpetuate domination" (36). Building community can extend beyond human-to-human connection to community building with the land itself, something Western paradigms of domination and extraction have socialized many people in the United States not to consider (Deloria and Wildcat). Like the best of partnerships, one with the land and plants is incremental, iterative, and evolving, based on respect, collaboration, imagination, and love.

"What's really key," Tim explains, "and what we are speaking about here—what we're offering to the community through this project—is an experience of tending the land that once was common in the human experience, and that only recently is absent from most of our daily lives. For countless millennia, no matter what part of the world your ancestors are from, they relied every day on a profound, living relationship with the land that provided for their needs. And I think that in our modern culture there has been a real drive to get away from that reality—that can at times be harsh—and toward a more easeful, comfortable existence. But in doing so, we become disconnected from the *truth* that we are dependent on an ongoing relationship with mother nature and all of her unpredictable outcomes. And when in times past we were more conscious of this relationship and immersed in it daily, with that came a profound sense of meaning and purpose in the world, and a profound appreciation and gratitude for the reality of the situation that we're in as human beings dependent on this natural world and in fact part of it. That unfortunately has today been mostly lost, with our pursuit of all of the comforts of modern, so-called civilized society. We are distancing ourselves more and more from our daily interdependence and relationship with the natural world."

Kerry, who has been nodding in agreement, adds, "[T]he cost has been great, perhaps greater than we can even really understand. And the loss has been greater than we really understand. And when we're here on this land, we begin to recognize what's been lost, and we feel the connection again."

"We've had the wool pulled over our eyes," Tim says, "and we've created this dark period of disconnection that permeates the human experience today. And the truth is that that very disconnection is itself an illusion. We human beings have always been inextricably part of the natural world and not separate from it, but we have constructed this elaborate ruse—this story that we are separate—and we are now so deeply immersed in that delusion that we feel it to be true. Climate change and pollution and loss of wild places are all a shared global suffering that has arisen out of a deluded refusal to recognize our profound interconnectedness with nature. And so, even in the very problems we now face as a result of our self-imposed disconnection from nature, you can see—clear as day—evidence that the opposite is in fact true: we are inextricably connected—with the natural world, and with each other. And the sooner we realize this, the sooner we can get ourselves unstuck from a lot of the problematic patterns that we perpetuate to this day. There are so many complex problems that we are facing today, and we feel that many—if not most—of those problems have their root in this illusion of separation from the natural world. And for a lot of people who have come to this place, myself included, being here has awakened or deepened that feeling of connection. Actually, more than just a feeling—a lived experience." This conversation about disconnection and unrest is certainly common in Indigenous writings that warn of the catastrophe of human disconnection.

Tim's comments make me think about his recent blog post about the murders of George Floyd and Brianna Taylor and so many other Black, brown, and Indigenous people in the United States. Tim, like many activists in the fight for food justice and racial justice, ties this violence to historical violence and racism all the way back to land theft, genocide, and slavery. I ask whether Tim and Kerry see part of their mission as in some small way trying to combat that kind of violent history of agriculture rooted in white settler colonialism and racism.

After a thoughtful pause Tim replies, "That's a big ask. I don't think our project could ever hope to directly address the huge social and cultural issues that you're alluding to there. But what I do think is that we can in some way expose the root distortions of perspective that *led* to the development of those problems in the first place. And I think that is what I was internally trying to work with when I was writing that blog post. The very patterns that led to the violent theft of the lands that we now call the Americas, and the violent enslavement of peoples to work those lands, those patterns are not a thing of the past, but they are deeply ingrained in our society today, without many [white] Americans ever noticing it. We, as a society, have not come to terms with that past.

"Not only have we not healed the wounds or righted the injustices, but it's even worse than that: we haven't let go of that way of viewing the world—and viewing our place in it—that *allowed* for those injustices to happen in the first place. The enslavement of a people to tend the land, and the stories about racial identity and superiority that were concocted in order to justify that slavery, it is all rooted in the white-European-colonialist delusion of land as a thing to be possessed, nature as a resource to be exploited, and the tending of the land as an undesirable burden to be pawned off onto other people. All of those same patterns of thinking are commonplace in our society today, even among the most well-intentioned people."

Tim's statements about the persistent legacies of violence against Black people, from enslavement to police shootings and criminalization of Black bodies, link to the historical work of numerous Black scholars such as Michelle Alexander and Ruha Benjamin, and in Rhetoric and Writing Studies, to Ersula Ore's scholarship on lynching and the idea that we must break from "white time" that posits these racial violences as in the past. When Ore discusses how uncomfortable white people in the United States are with a comparison between Trayvon Martin's killing and Emmett Till's lynching, she explains,

> [S]uch narrow definitions for what constitutes a lynching permitted conservatives to deny the continuity between Trayvon and Emmett slayings and to conclude that those who read Trayvon and Emmett as "the same thing" were more interested in stocking the fires of racial divide than in encouraging racial unity. I recall the controversy over the Emmett-Trayvon comparison for the way it highlights continued debate over the definition of lynching and the ways such debate reflects an ongoing tradition to rhetorically save face through denial of lynching's adaptive and transformative nature. (*Lynching*, 4)

Ore continues, "To assert that the Trayvon is Emmett comparison lacks validity on account of form, motive, and character is to reject the changing same that is American racism" (5). In a related issue of continued violence against sovereign Indigenous Nations, Andrea Riley Mukavetz and Cindy Tekobbe write, "Even today there are continued efforts to remove what little rights and access we have left to our sacred lands and places, to sever the connections between us and our relations." Racist and settler colonial legacies at the founding and expansion of the United States are pervasive in most elements of US culture today.

When white Americans like me fail to acknowledge these legacies, we perpetuate the violence. When we do not recognize our current industrial food system as part of this legacy, we perpetuate the violence. When we do not recognize discussions of white agricultural heritage as continued settler colonialism, we perpetuate the violence. When we extract and destroy in the name of progress and unlimited growth rather than turn toward relationships with all fellow beings, we perpetuate the violence. The severing of humans and nature is an ongoing violence that many BIPOC scholars, organizers, and activists are trying to repair; in a different though related way, so are regenerative farmers. Food literacy practitioners can make the connections visible between regenerative and local food work and the fight against racial injustice and settler colonialism. When we do not explicitly name the latter in discussions of the former, we may perpetuate legacies of trauma, as some of the work happening in Boulder County and indeed some of the project artifacts I showed in the previous chapter demonstrate.

As Tim explains, "Part of what we're doing here with our project is sharing our deeply personal discovery that the human act of tending the land can give back in profound ways to the person who engages in that relationship. It can alter our perception of our place in the world. It can alter our understanding of our relationship with nature, and our relationship with our neighbors."

Tim encapsulates the reasons for why there is an urgency to expand our community partnership paradigm. The ecological catastrophe is also a human and spiritual catastrophe. It continues today in every facet of our lives. The industrial food system depends on and only works because of exploitation of poor workers, mostly Black and brown, the migrant farmers, the meat packers, the factory workers. Sometimes there is literal slavery, as was exposed in Florida with tomato farmers (Estabrook). The exploitation and racial, colonial violence persist. The destruction of land and destruction of humans are inextricably connected.

Kerry agrees, "There are complex, systemic problems that will require complex, systemic solutions.... Seeing land as a possession, seeing nature as a resource to be extracted, seeing work on the land as menial labor—these mindsets contribute to and even perpetuate the other complex problems we have now. And if we can address that root relationship—'right relationship' with the natural world—then I feel we have a better chance of moving in the direction of healing."

In trying continually to build accountability into our work, the other-than-humans write and unwrite, teaching us, partnering with us, collaboratively healing and offering. Can writing occur in the form of a vibrant carrot or a creek that cuts a property or a patch of grass or the grazing of undomesticated animals on uncultivated land? Here we are, sitting at the picnic table, drinking tea made from the garden's herbs, gazing at the grass growing from soil filled with bustling microbes and bugs, and the birds cocking their heads looking for grubs, and the deer drinking from the bird bath nearby, and the tall grasses swaying in the wind. In this mix of human and other-than-human, we listen as their rhetorical lives write the answer before us.

Critical Food Literacy Through Relational Abundance

The ethical, relational work that Kerry and Tim practice connects to a larger educational endeavor for food literacy and justice activists and advocates inside and outside of academia. Tim and Kerry share with their community the literacies that arise from partnership with the land as they counter some of the injustices of the industrial food system and the individualism and models of scarcity and mindless accumulation perpetuated by white supremacist and capitalist ideologies. Tim argues, "[I]t's because of the absence of an authentic relationship with the land, and with the plants and animals we eat, that the conditions allowed for an industrial food system to develop. If we are to dismantle that industrial food system and reimagine what it means to have healthy food, healthy planet, healthy societies, it all starts with first having a relationship with the land."

Rhetoric and Writing Studies teachers, scholars, and students have a vital role, as part of a food justice framework, to draw connections between partnering with land in regenerative production and dismantling oppression in the food system—these go hand in hand. As Iris D. Ruiz explains in her work to reclaim composition studies for Chicanx and other minoritized groups, we must continuously work to recover "excluded histories" (*Reclaiming* 150). She explains, "[W]hen particular populations are left out of texts thought to represent a foundational understanding of a field, then the needs of those populations will also be marginalized" (34–35).

Ecological community writing is not simply a methodology for expanding the ways we work and think as an intellectual endeavor; it is a means of disruption and survival, a reconnection with and accountability to legacies of the past and with a future that we co-imagine. Ruiz urges us to recover, shatter, and disrupt histories (151). *Ecological community writing* calls for a reorienting to land, other humans and beings, affect, and language to prioritize relationships and connection as well as to acknowledge in a deeply reflective way the traumas that persist. That needs to happen here in Boulder County and in every community throughout the country, and it needs to occur personally in many of us. An *ecological community writing* methodology works toward dismantling systems of oppression because it does not shy away from systems thinking, relational work, and personal and community reckoning with legacies of the past that continue to do harm.

In food literacy work, this means shattering and disrupting settler colonial histories of acquisition and paradigms of excess such as "bigger, fatter, faster, cheaper" (*Food, Inc.*), while recovering paradigms of connection, community accountability, and relationship. Through classes, workshops, and community events that connect people with land and plants and seeds, Kerry and Tim share the land's messages and history with others. Reviving people's profound reciprocal relationship with land and food is a radical act connecting local food system work to food justice work. Tim and Kerry's pay-as-you-can and take-what-you-need model shifts away from a transactional approach to a transformational one that foregrounds relationships and community. People learn on and from the land, financially or physically support the land if they can, and, each week, accept the land's gifts in return. This is a food access model outside of industrial or corporate capitalist frameworks, and people's participation in this model helps to "write" the possibilities for how a community defines its local food system and who gets to eat healthy food. Tim and Kerry's model is part of the *distributed definition building* happening across the county. They and the land are co-writing a definition of "local food."

When we enter into partnerships with farmers, government officials, food organizers and advocates, and others through our scholarship and teaching, they can work with us on developing projects and curricula. When we turn the typical triangle of reciprocity between human teacher—human student—human community project partner into an ecology that includes other-than-humans in intentional design, we can

FIGURE 6.2. After researching regenerative agriculture, writing students spend a day at Dharma's Garden learning from founding farmers Tim Francis and Kerry Francis. Here they are planting asparagus at the farm. (Photo credit: Veronica House)

build into our course projects and readings for students an attention to the authorial power of other-than-humans as collaborators, encouraging a food literacy that attunes to all elements of the process of food production and distribution and to the ecology of relationships among all involved.

When my writing students would take trips to Dharma's Garden, we began with a contemplative writing exercise. They would walk the land or find a spot to sit without talking or using phones, and they would write

FIGURE 6.3. Students weed one of the garden rows at Dharma's Garden. (Photo credit: Veronica House)

down all that came to them—observations of the animals or plants; sensory observations; what feelings surfaced in their bodies; and any other thoughts they wished to jot down. After learning from Kerry and Tim throughout the day, hearing about the history of the land that is now Dharma's Garden and their philosophy of farming with the land, and after engaging in a planting or harvesting task that connects the students, the land, and the plants, they would write again based on a prompt from me or from Kerry and Tim, urging them to make connections between broader issues and readings we'd studied and their emotional and physical learnings on the land (see figures 6.2–6.3). Often, their perspective deepened to some degree, marking an appreciation and surprise for such beauty and abundance in the heart of the city and gratitude for the feelings that arose in them as they felt the land beneath their feet and in their hands. Back in the classroom, I would invite the students to reflect again to connect their contemplative writings from the farm to any feelings of grief, anger, joy, longing, questioning, and other complex emotions that may have come up during or after the visit.

When we think of other-than-humans as partners in our work and as part of the community we love, the focus changes and expands. As Madeline, my former writing student, said so beautifully in our interview, "visiting the farm [in our class] was completely transformative because I got to see it in action, particularly because Dharma's Garden is so special, it's, you know, just so wild and magical and abundant. You can tell that the land is happy. It's reflected by the vibrant veggies and flowers themselves, which served as my personal entry point into this relationship [see figure 6.4]. If I didn't have the opportunity to actually put my hands in the dirt there, it totally would have been different. No matter how many documentaries we could have watched or articles we read, that hands-on learning is huge" (Nall).

In their article on decolonizing community writing through community listening, Rachel Jackson and Dorothy Whitehorse DeLaune ask scholars to actively resist the settler colonial force of Western academic moves to quantify and to logically "settle" meaning. In a decolonial Kiowa approach, for example, meaning making happens through "co-construction of narrative and communal articulation of meaning" (43). Non-Indigenous scholars can learn a tremendous lesson here and from Riley Mukavetz and Tekobbe, who argue that "to be a good Indigenous scholar in higher education . . . is to navigate your [white scholars'] western knowledge systems that 'identify,' count, and categorize in an effort to fix knowledge in the euro-centric tradition of objectivity when we know knowledge is negotiated between people within a context of relations and collective identity." As a non-Indigenous scholar learning from this, how can I continuously challenge and question top-down knowledge systems created and justified by people in power? When my partners and I encourage distributed community writing about what "local food" is, we do not attempt to "settle" meaning. Rather, we understand that we are all continuously co-constructing in relationship in

FIGURE 6.4. Dharma's Garden farm stand displays vibrant colors of produce. (Photo credit: Natasha S. Hansen, PhD)

order to shift some of the definition-building power to community members, including other-than-humans.

In a related way, this co-constructed and unsettled nature of meaning gets at the pre-language conversations Tim and Kerry describe having with the land and plants. While they can't easily articulate how the conversation reveals itself, they know it is occurring through their co-constructed rhetorical embodiment of their collaborative desires. It is akin to what Leanne Betasamosake Simpson has called "land as pedagogy," learning *"from* the land and *with* the land" and sharing with our students that we are "dependent upon intimate relationships of reciprocity, humility, honesty and respect with all elements of creation, including plants and animals" (7–10, original italics). Many Indigenous cultures across the world share similar concepts. Simpson's idea recalls Shane Bernardo and Terese Guinsatao Monberg's discussion of the Indigenous Filipinx concept of *kapwa*, in which "If I harm you, I harm myself. If I love you, I love myself. There is a quality of intimacy and shared risk involved. We are not separate. We are connected" (87). Reciprocity binds us in deep intimacy and partnership across our communities, and this distributed kind of reciprocity illuminates definitions for what local food can be—definitions that might never have been revealed through a partnership model that privileges humans only.

To collaboratively write meaning in partnership with other-than-humans can be a curative act. Iris D. Ruiz posits a concept of "medicinal rhetorics" through the lens of Curanderismo, an Indigenous healing praxis "in Mexico [that] is based on Aztec, Mayan, and Spanish influences" and "serves as a type of 'medicinal' practice that cures and revives while it reclaims the practices which were already in existence before colonization" (Lockett et al. 45, 58). She calls her epistemic practice "historical curanderisma," which she describes as an ongoing "delinking from dominant disciplinary discourses of whiteness associated with" rhetoric, composition, and writing studies (Lockett et al. 45). "Historical Curanderisma," Ruiz explains, "informs my practice to better confront the epistemic trauma often expressed by 'colonized scholars,' who do not see themselves in claimed critical methodologies" (Lockett et al. 54). Ruiz explicitly connects the epistemic violence to dispossession from land:

> An embodied rhetorical practice that recognizes the ways Chicanxs are forced to deny the histories of their own bodies could be one step toward encouraging a decolonial theory of embodiment that looks at the body of the indigenous as one that has been colonized, traumatized, divorced from their own land, and de-legitimized in the name of both capitalism and a democracy that values profit over people and schooling systems that support a neoliberal political agenda. An embodied rhetorical practice is a praxis that acknowledges this reality for minoritized populations.... It is an engagement with Curandera praxis as healing praxis. (Lockett et al. 63)

Ruiz's development of her methodological practice of "historical curanderisma" is "a survival mechanism for the colonized mind, body, and spirit that is rendered invisible in academia" (Lockett et al. 69). As Ruiz is an Indigenous Chicana scholar, her methodological practice serves different purposes than my own learning from her practice does. However, in respect for her vision I ask myself what my obligation is as a descendant of white settler colonizers here on this land where I live, to work toward healing myself, to whatever extent possible, of my own Eurocentric ways of knowing that are so deeply embedded in my consciousness and, as Menakem writes, into my very DNA. I do not want to appropriate Ruiz's term, which is designated, as she says, for "minoritized populations." As a non-Indigenous scholar, I must engage in a different but related lifelong process of healing the parts of me that have been made sick through internalization of colonial paradigms.

Settler colonialism and white supremacy culture are deeply woven into the industrial food system; they are woven into disconnection and dispossession from the land; and they are woven into our educational system and structures in higher education. These truths call me to hold myself accountable to the pain I inflict on other humans, animals, and land. They also call me to create and experience joy and connection with humans, animals, and land. They call me to do whatever I can to challenge educational and research systems that accept certain knowledges and ignore or denigrate others. They call on me to challenge the violence of food systems that reinforce racist, colonial, ableist frameworks that place wealthy, white humans at the top; disregard the creative potential of other-than-humans; and allow for extraction, exploitation, and scarcity ideologies to persist. They call me to reconceive community writing through an ecological methodology that reimagines my work as part of an ecology and as helping to build an ecology that accounts for language, affect, humans, organizations, other-than-humans, and land.

The Personal Work Is Systemic Work, a Closing and an Opening

In January 2021, three days before the first day of spring semester classes and, as it turned out, my last semester at University of Colorado Boulder before I moved to a position at University of Denver, I was reassigned from my regular upper-division Food and Culture writing courses to an online synchronous first-year writing class. I hadn't taught first-year writing in a decade, but I adapted my class to fit the course goals and hoped that some of my partners would visit our class remotely. Even if, because of COVID-19, students couldn't have the hands-on experiences that are a staple of my in-person classes, they could still learn from people in the community who visited the class via Zoom. As with my in-person classes, we looked at previous students' projects so that my current students could understand their writing in a larger context of writing happening across our community and across time. An *ecological community writing* methodology is adaptable to classroom-only and online teaching.

During a discussion of factory farming and grass-based ranching, a white, male student shared his experience as part of a ranching family that lives on land that was one of the original settler homesteads in Boulder County. His family had raised cattle there for three generations. His grandfather, in his nineties, had lived on the property for most of his life. After class, I asked my student whether he could ask his grandfather if he knew the Gould homestead location. I had spent several years trying to locate the original homestead where my great-great-grandmother Marinda had lived as a child. I knew the general location, but historical records and archives I searched hadn't been able to provide the exact spot. Before our next class began, my student joined Zoom early, eager to share his grandfather's news: "We live on that land! We live on the old Gould farm! My grandfather remembers the older Goulds from when he was a boy. But we're selling the property. It's getting to be too much work." He gave me the address so I could look up the property. That I happened to be reassigned this class section out of hundreds with this particular student out of thousands is a coincidence beyond my comprehension.

After the course ended, my family and I toured the land. Before it sold, I was able to walk the fields with my two daughters (see figure 6.5). My youngest was the same age as Marinda was when she first arrived here. I touched the old wood door on the milk barn, one of the original homestead's outbuildings, and imagined Marinda's hand touching that spot. I tried to imagine her beside me as I walked across the meadow. I asked her and the land itself over and over what I'm here in Colorado to do. Why have I been drawn to this ancestral spot, first through my job, now through my student? What memories and traces of joys and violences does the land hold, and does my body hold? What traces of Marinda are bound in my DNA? And what does a homesteader's great-great-granddaughter do to contend with the complexities of this side of my family history? Marinda had written that farming was "the hardest kind of hard work." What, as a teacher, scholar, and descendant of colonizers, is my hardest kind of hard work? As Qwo-Li Driskill reminds us, "[i]t is our responsibility as both Native and non-Native people living on occupied land to disrupt colonial projects through our work" (76). As a lifelong, recursive question, I need to ask what that work should be and then continually "do the work," as Carmen Kynard would say.

An *ecological community writing* methodology calls each of us to not only work for justice in our communities but also in our own legacies. I can't

FIGURE 6.5. This pasture was part of the original Gould homestead in Boulder. (Photo credit: Veronica House)

know that I have settler colonial ancestors and, as Ersula Ore suggests, think that the violences they inflicted are in the past. I have to see the work that needs to happen around how rhetoric is used in the county and city now as part of my ancestral legacy. I can't see the personal work I need to do as separate from the community writing work I facilitate or as separate from the writing work my students and partners do. It is all connected because we are all part of a connected ecology.

Ecological community writing moves students and teachers between a study of global, national, and local discourses and definitions of contested terms. It challenges us to generate writing, events, and projects with a broad interdisciplinary, intercommunity, and interspecies focus on *distributed definition building*. And it asks us to continually do the internal work that helps us make sense of our relationship to our particular locale. The assignments, events, and projects we develop through *ecological community writing* will look different depending on where we live and what our positionality is in the world. We can begin by opening ourselves to the calling we feel from the place where we live and by listening to its teachings. Listening to the land and what it has to teach us, give us, and ask of us brings to mind Kimmerer's discussion of the Honorable Harvest (*Braiding* 183). Ask permission first. Only take what you need. Give a gift in return.

For teachers, what might these three steps of asking permission, only taking what we need, and giving a gift in return look like? As I try to live into these calls, I offer a partial answer in terms of how I am adjusting my courses at the University of Denver. My students and I are studying and building justice-oriented literacy projects around the lands' official and contested histories (Ruiz 181). We are studying writings that teach of sacred connection and kinship between species and things. I center and financially compensate local BIPOC and other frontline food justice advocates, activists, organizers, farmers, and scholars who are working to dismantle systems of oppression. I incorporate several outdoor classes to give students experience at urban farms and gardens in our community that connect students' learning to the land and animals we study. I encourage students to build relationships and an understanding of the systemic, ecological nature of issues and their localized particularities through readings, research, and physical connection.[2] I ask students to investigate current Native-run food sovereignty programs. We study Colorado food cultures by connecting with and having students participate at events at farms like the BIPOC-led Frontline Farming. Through these connections to past and present, my students and I are continually developing awareness of multiple knowledges

---

2 Joyce Rain Anderson offers wonderful concrete ideas for substantive pedagogical and institutional commitments in "Remapping Settler Colonial Territories."

and an understanding of Rhetoric and Writing's potential as embodied and distributed practices across humans and other beings. If we listen closely, the place calls forth the curriculum in partnership with us.

For a scholar, what do the three steps of the Honorable Harvest look like? Again, in my ongoing process of learning, it has meant asking the land, and those who have built deep relationships with the land, what is needed. My human partners are my teachers, too. The farmers and activists I partner with can often translate for me, teaching me the languages I haven't yet learned. As I work with an *ecological community writing* methodology, I continuously try to bring in and seek multiple knowledges, not all scholarly, and not all human. I come to a project without a set agenda, opening myself to an evolving research process that shifts and transforms continuously along with my human and other-than-human partners' teachings, histories, questions, and needs. It means offering BIPOC-led organizations and initiatives my skills and resources, when invited. It means acknowledging my own limited perspective and pledging to continually learn more and do better to help resist and shift power imbalances.

As we attune to the animate grammars around us, they can alter our decisions; enhance our scholarly, organizing, and activist work; and transform our curricula. To be sure, compared to the abundance offered to us in partnership, ours are small gestures of reciprocal gift giving, but they are a beginning, a seed for something to come.

How is abundant *ecological community writing* local and organic, as this book's title suggests? We can reimagine Rhetoric and Writing Studies' work with community literacies and engagement in ways that are locally responsive and that emerge organically across a broad range of community members, researchers, teachers, organizations, other-than-humans, students, language choices, and histories to encourage collaboration, reciprocity, and co-creation of knowledge. When we, our students, and our project partners understand the ecological nature and complexity of our work, then the writing we produce is both part of *and is creating* connections in a continuously evolving, multifaceted, abundant ecology always in the process of being written.

Works Cited

Abisaid, Joseph L. "Tying the Knot: How Industry and Animal Advocacy Organizations Market Language as Humane." *The Political Language of Food*, edited by Samuel Boerboom. Lexington Books, 2015, 141–53.

Abram, David. *The Spell of the Sensuous: Perception and Language in a More-Than-Human World*. Vintage Books, 1997.

Acosta, Angel. "Contemplative Science and Practice Through a Healing-Centered Perspective." Keynote address. Center for Contemplative Mind in Society 5 Aug. 2020. Online.

Adams, Jennifer L. "Family Farms with Happy Cows: A Narrative Analysis of Horizon Organic Dairy Packaging Labels." *The Political Language of Food*, edited by Samuel Boerboom, Lexington Books, 2015, pp. 183–200.

Adler-Kassner, Linda. *The Activist WPA: Changing Stories about Writing and Writers*. Utah State UP, 2008.

Ahmed, Sara. "The Nonperformativity of Antiracism." *Meridians*, vol. 7, no. 1, 2006, pp. 104–26. https://www.jstor.org/stable/40338719.

Alexander, Michelle. *The New Jim Crow: Mass Incarceration in the Age of Colorblindness*. The New Press, 2010.

Alkon, Alison Hope, and Julian Agyeman, editors. *Cultivating Food Justice: Race, Class, and Sustainability*. MIT P, 2011.

Alkon, Alison Hope, and Julie Guthman, editors. *The New Food Activism: Opposition, Cooperation, and Collective Action.* U of California P, 2017.

Aloziem, Ozy. "A Dinner Party Game to Spark Your Radical Imagination." *TEDxMileHigh*, Denver, Oct. 2021, https://www.youtube.com/watch?v=sTbMjVqoWxs.

Alvarez, Steven. *Brokering Tareas: Mexican Immigrant Families Translanguaging Homework Literacies.* State U of New York, 2017.

Alvarez, Steven. "Taco Literacies: Ethnography, Foodways, and Emotions Through Mexican Food Writing." *Composition Forum*, vol. 34, Summer 2016, https://www.compositionforum.com/issue/34/taco-literacies.php.

Anderson, Joyce Rain. "Remapping Settler Colonial Territories: Bringing Local Native Knowledge into the Classroom." *Survivance, Sovereignty, and Story: Teaching American Indian Rhetorics*, edited by Lisa King, Rose Gubele, and Joyce Rain Anderson, Utah State UP, 2015, pp. 160–69.

Arellano, Sonia C., Lauren Brentnell, Jo Hsu, and Alexis McGee. "Ethically Working Within Communities: Cultural Rhetorics Methodologies Principles." *constellations*, vol. 4, June 2021, https://constell8cr.com/conversations/cultural-rhetorics-methodologies/.

Arthur, Mikaila Mariel Lemonik. "Qualitative Coding." *Social Data Analysis.* 2021. Web. https://pressbooks.ric.edu/socialdataanalysis/chapter/qualitative-coding-2/. Accessed 31 Dec. 2024.

Bacon, Nora. "The Trouble with Transfer: Lessons from a Study of Community Service Writing." *Michigan Journal of Community Service Learning*, vol. 6, no. 1, 1999, pp. 53–62.

Baker-Bell, April. *Linguistic Justice: Black Language, Literacy, Identity, and Pedagogy.* NCTE-Routledge, 2020.

Barth, Brian. 2016. "Farmers Are Capitalizing on Carbon Sequestration: How Much Is Your Carbon-Rich Soil Worth?" https://modernfarmer.com/2016/04/carbon-sequestration/. Accessed 25 May 2018.

Bartlett, Al. "Arithmetic, Population, and Energy." *Global Public Media*, 26 Feb. 2005. DVD.

Bauer, Holly. *Food Matters: A Bedford Spotlight Reader.* Bedford/St. Martin's, 2014.

Bayer Group website. https://www.cropscience.bayer.com. Accessed 20 Apr. 2023.

Benjamin, Ruha. *Viral Justice: How We Grow the World We Want.* Princeton UP, 2022.

Bennett, Jane. *Vibrant Matter: A Political Ecology of Things.* Duke UP, 2010.

Berger, Jonah. *Contagious: Why Things Catch On.* Simon and Schuster Paperbacks, 2013.

Bernardo, Shane, and Terese Guinsatao Monberg. "Resituating Reciprocity Within Longer Legacies of Colonization: A Conversation," *Community Literacy Journal*, 2021, pp. 83–93, https://doi.org/10.25148/clj.14.1.009058.

BestPlaces. "2021 Cost of Living Calculator Boulder CO." https://www.bestplaces.net/cost_of_living/city/colorado/boulder. Accessed May 2022.

Blackburn, Lorelei, and Ellen Cushman. "Assessing Sustainability: The Class That Went Terribly Wrong." *Unsustainable: Re-imagining Community Literacy, Public Writing, Service-Learning, and the University*, edited by Jessica Restaino and Laurie JC Cella, Lexington Books, 2013, pp. 161–77.

Blanco-Wells, Gustavo. "Ecologies of Repair: A Post-Human Approach to Other-Than-Human Natures." *Frontiers in Psychology*, vol. 12, 2021, https://doi.org/10.3389/fpsyg.2021.633737.

Bland, Alastair. "A Legal Twist In the Effort to Ban Cameras from Livestock Plants." npr.org, 11 Apr. 2013, https://www.npr.org/sections/thesalt/2013/04/10/176843210/a-legal-twist-in-the-effort-to-ban-cameras-from-livestock-plants.

Bloom, Lynn Z. "Consuming Prose: The Delectable Rhetoric of Food Writing." *College English*, vol. 70, no. 4, 2008, pp. 346–61.

Boerboom, Samuel, editor. *The Political Language of Food*. Lexington Books, 2015.

Booth, Michael. "Boulder County Allowing Farmers to Grow GMOs on Open Space After Organic Utopia Didn't Materialize." *Colorado Sun*, 7 Jan. 2022. Web. https://coloradosun.com/2022/01/06/boulder-county-removes-gmo-ban/.

Born, Branden, and Mark Purcell. "Avoiding the Local Trap: Scale and Food Systems in Planning Research." *Journal of Planning Education and Research*, vol. 26, no. 2, 1 Dec. 2006, pp. 195–207.

Boulder County. "Agriculture Statistics, Acres, and Overview." 6 Nov. 2015. Boulder County Comprehensive Plan, https://www.bouldercounty.org/property-and-land/land-use/planning/boulder-county-comprehensive-plan/.

Boulder County. "Boulder County Comprehensive Plan." Adopted 15 July 2020. https://assets.bouldercounty.gov/wp-content/uploads/2018/10/bccp-boulder-county-comprehensive-plan.pdf.

Boulder County. "Boulder County Land Use Code." Edited by Boulder County Land Use Department. Boulder, 2021.

Boulder County. "Boulder County Parks and Open Space Cropland Policy." 2021. https://assets.bouldercounty.gov/wp-content/uploads/2021/12/cropland-policy.pdf.

Boulder County. "Exploring Solutions to Boulder County Agricultural Sector Constraints." Aug. 2022. https://assets.bouldercounty.gov/wp-content/uploads/2022/09/ag-sector-contraints-findings-recommendations.pdf.

Boulder County. "Invasive Plants and Weed Management on Open Space." https://bouldercounty.gov/open-space/management/weeds/. Accessed 8 June 2024.

Boulder County. "Local Food and Sustainable Agriculture." https://bouldercounty.gov/environment/sustainability/food-and-agriculture/. Accessed 6 Nov. 2015.

Boulder County. "Restore Colorado." https://assets.bouldercounty.gov/wp-content/uploads/2023/07/BoCo_RestoreReport_06-1.pdf. Accessed 18 Dec. 2024.

Boulder County. "September 2022 Consultation Statement." https://bouldercolorado.gov/september-2022-consultation-statement. Accessed 21 Oct. 2024.

Boulder County Open Space. "Parks and Open Space Founders' Legacy." *YouTube*, 9 May 2017, https://www.youtube.com/watch?v=vuHQs5riG9c.

Boulder Food Rescue. Website. https://www.boulderfoodrescue.org. Accessed 10 May 2021.

Boulder Food Rescue. *2021 Annual Report*. https://www.boulderfoodrescue.org/wp-content/uploads/2022/03/BFR-2021-annual-report-website.pdf.

Brain, Roslynn. "The Local Food Movement: Definitions, Benefits, and Resources." *Utah State University Extension Sustainability*, Sept. 2012, pp. 1–4, https://digitalcommons.usu.edu/cgi/viewcontent.cgi?article=2693&context=extension_curall.

Brandt, Deborah. *Literacy in American Lives*. Cambridge UP, 2001.

Brewster, Cori. "Toward a Critical Agricultural Literacy." *Reclaiming the Rural: Essays on Literacy, Rhetoric, and Pedagogy*, edited by Kim Donehower, Charlotte Hogg, and Eileen E. Schell, Southern Illinois UP, 2012, pp. 34–51.

Broad, Garrett M. *More Than Just Food: Food Justice and Community Change*. U of California P, 2016.

Brooke, Collin. *Lingua Fracta: Towards a Rhetoric of New Media*. Hampton Press, 2009. Print.

brown, adrienne maree. "The Darwin Variant, and/or Love of the Fittest." 19 Aug. 2021. Blog.

brown, adrienne maree. *Emergent Strategy*. AK Press, 2017.

brown, adrienne maree. "an emergent strategy response to mass shootings." 7 June 2022, https://adriennemareebrown.net/2022/06/07/an-emergent-strategy-response-to-mass-shootings/.

brown, adrienne maree. *Holding Change: The Way of Emergent Strategy Facilitation and Mediation*. AK Press, 2021.

brown, adrienne maree. "intersecting worlds: the one we've got, the one we're building, the ones we imagine," plenary presentation notes. 8 June 2014. https://adriennemareebrown.net/2014/06/08/my-talk-notes-from-commonbound-intersecting-worlds-the-one-weve-got-the-one-were-building-the-ones-we-imagine/.

brown, adrienne maree. *Pleasure Activism: The Politics of Feeling Good*. AK Press, 2019.

brown, adrienne maree, guest. "On Radical Imagination and Moving Towards Life." *On Being*. 23 June 2022. https://onbeing.org/programs/adrienne-maree-brown-on-radical-imagination-and-moving-towards-life/.

Brownlee, Michael. "Increasing Food Production Capacity." Public Address to Local Food Think Tank, 28 July 2014, Denver.

Brownlee, Michael. "The Local Food and Farming Revolution." *Resilience*. 9 Mar. 2010. https://www.resilience.org/stories/2010-03-09/local-food-and-farming-revolution/. Accessed 14 Dec. 2024.

Brownlee, Michael. *Sustainovation Keynote*. 9 Apr. 2014, Denver, CO.

Brownlee, Michael. "Thinking Like a Foodshed." *LocalFoodShift.com*, 2012.

Carter, Shannon. "Pass the Baton: Racial Justice: Lessons from Historic Examples of the Political Turn 1967–68." *Writing Democracy: The Political Turn in and Beyond the Trump Era*, edited by Shannon Carter, Deborah Mutnick, Jess Pauszek, and Steve Parks, Routledge Press, Aug. 2019.

Carter, Shannon, and Deborah Mutnick. "Writing Democracy: Notes on a Federal Writers' Project for the Twenty-First Century." *Community Literacy Journal*, vol. 7, no. 1, 2012, pp. 1–13.

Castle, Shay. 2016. "80 Percent Failure Rate Dogs County's $1 Million Organic Farm Program." *Daily Camera*, last modified 12 Mar., http://www.dailycamera.com/boulder-business/ci_29627753/80-percent-failure-rate-dogs-countys-1-million.

Cedillo, Christina. "Unruly Borders, Bodies, and Blood: Mexican 'Mongrels' and the Eugenics of Empire." *Journal for the History of Rhetoric*, vol. 24, no. 1, 2021, pp. 7–23. https://doi.org/10.1080/26878003.2021.1881307.

Center for Food Safety. "About Genetically Engineered Foods." 2025. https://www.centerforfoodsafety.org/issues/311/ge-foods/about-ge-foods.

City of Boulder. "Boulder's Open Space and Mountain Parks: Then and Now." 2008. https://www-static.bouldercolorado.gov/docs/boulders-open-space-and-mountain-parks-then-and-now-1-201401291109.pdf. Accessed 2 Feb. 2018.

City of Boulder. "Fort Chambers and the Sand Creek Massacre." https://bouldercolorado.gov/fort-chambers-and-sand-creek-massacre. Accessed 3 Aug. 2022.

City of Boulder. "Indigenous Peoples Projects." https://bouldercolorado.gov/services/indigenous-peoples-projects. Accessed 27 July 2024.

City of Boulder. "Staff Land Acknowledgment." https://bouldercolorado.gov/projects/staff-land-acknowledgment. Accessed 27 July 2024.

City of Boulder. "Tribal Nation Agreements." https://bouldercolorado.gov/city-tribal-nation-agreements. Accessed 3 Aug. 2022.

"Climate Change 2014. Impacts, Adaptation, and Vulnerability." UN Intergovernmental Panel on Climate Change. Web. 5 Nov. 2015.

Coalition for Community Writing. "Promotion and Tenure Resources." https://communitywriting.org/promotion-and-tenure-resources/. Accessed 21 Oct. 2024.

Coleman, Eliot. "Beyond Organic." *Food*, edited by Brooke Rollins and Lee Bauknight, Fountainhead, 2010, pp. 113–16.

Coleman-Jensen, Alisha, and Mark Nord. "Disability Is an Important Risk Factor for Food Insecurity." *USDA Economic Research Service*, 6 May 2013. Web. https://www.ers.usda.gov/amber-waves/2013/may/disability-is-an-important-risk-factor-for-food-insecurity/.

Conley, Donovan, and Justin Eckstein. *Cookery: Food Rhetorics and Social Production*. U of Alabama P, 2020.

Cooks, Leda. "Constructing Taste and Waste as Habitus: Food and Matters of Accessibility In/Security." *The Political Language of Food*, edited by Samuel Boerboom, Lexington Books, 2015, pp. 123–40.

Cooper, Ann. Interview. By Veronica House, 10 June 2020.

Cooper, Marylin. "The Ecology of Writing." *College English*, vol. 48, no. 4, 1986, pp. 364–75.

Cooper, Marylin. "Foreword." *ecocomposition*, edited by Christian R. Weisser and Sidney I. Dobrin, SUNY P, 2001, pp. xi–xviii.

Coppom, Brian. "Interview with Boulder County Farmers Markets Executive Director Brian Coppom." *Local Food Shift Colorado* 1, 2015, pp. 80–86.

Coppom, Brian. Interview. By Veronica House, 12 June 2020.

Cullen, Art. "Politicians Say Nothing, but US Farmers Are Increasingly Terrified by It—Climate Change." *The Guardian*. Friday 19 Oct. 2018, https://www.theguardian.com/environment/2018/oct/19/politicians-say-nothing-but-us-farmers-are-increasingly-terrified-by-it-climate-change?CMP=Share_iOSApp_Other&fbclid=IwAR2rsIDgvYG0V6IaduQ6jvJ7AMmgOLN3NTdSoH9r58r23ba2IDd1WKX6EwU.

Cushman, Ellen. *The Cherokee Syllabary: Writing the People's Perseverance*. U of Oklahoma P, 2011.

Cushman, Ellen. "The Rhetorician as an Agent of Social Change." *College Composition and Communication*, vol. 47, no. 1, 1996, pp. 7–28.

Dancer, Ashley. *How Can Public Policy Create Opportunities and Barriers to the Development of a Local Food System? An Example of How Land-Use Policies Affect Small, Local Farmers in Boulder County*. 2018. Leeds School of Business and Environmental Studies Program, U of Colorado, MA thesis.

Dancer, Ashley, Peter Newton, and Veronica House. "Perceived Opportunities for, and Barriers to, the Development of Local Food Systems: A Case Study from Boulder County, Colorado." *Food Studies: An Interdisciplinary Journal*, vol. 9, no. 4, 2019, pp. 1–20.

Danovich, Tove. "How a Seed Bank Helps Preserve Cherokee Culture Through Traditional Foods." *npr online*. April 2, 2019. https://www.npr.org/sections/thesalt/2019/04/02/704795157/how-a-seed-bank-helps-preserve-cherokee-culture-through-traditional-foods.

Davis, Angela. *Abolition and the Radical Imagination*. Keynote. Critical Resistance, 20 Feb. 2016, Culver City, CA.

Davis, Angela. *Politics and Aesthetics in the Era of Black Lives Matter*. Lecture. NYU Skirball Talks, 5 Nov. 2018.

Deans, Thomas. "English Studies and Public Service." *Writing and Community Engagement: A Critical Sourcebook*, edited by Thomas Deans, Barbara Roswell, and Adrian J. Wurr, Bedford/St. Martin's, 2010, pp. 97–116.

Deans, Thomas. "Sustainability Deferred: The Conflicting Logics of Career Advancement and Community Engagement." *Unsustainable: Re-Imagining Community Literacy, Public Writing, Service-Learning, and the University*, edited by Jessica Restaino and Laurie J. C. Cella, Lexington Books, 2013, pp. 101–11.

Deans, Thomas. *Writing Partnerships: Service-Learning in Composition*. National Council of Teachers of English, 2000.

Del Hierro, Victor, Valente Francisco Saenz, Laura Gonzales, Lucia Durá, and William Medina-Jerez. "Nutrition, Health, and Wellness at La Escuelita: A Community-Driven Effort toward Food and Environmental Justice." *Community Literacy Journal*, vol. 14, no. 1, 2019, pp. 26–43.

Deloria, Vine, Jr., and Daniel Wildcat. *Power and Place: Indian Education in America*. Fulcrum Publishing, 2001.

Dharma's Garden. Website. https://www.wonderhaven.org.

Dingo, Rebecca. *Networking Arguments: Rhetoric, Transnational Feminism, and Public Policy Writing*. U of Pittsburgh P, 2012.

DJ Cavem. "DJ Cavem Bio." *DJ Cavem*. https://www.chefietef.com/dj-cavem-bio/.

Dobie, Kathy. "This Is Not Farming." *O Magazine*, Nov. 2011, pp. 172–209.

Dobrin, Sidney I., editor. *Ecology, Writing Theory, and New Media*. Routledge, 2012.

Dobrin, Sidney I. *Postcomposition*. Southern Illinois UP, 2011.

Dobrin, Sidney I., and Christian R. Weisser. *Natural Discourse: Toward Ecocomposition*. SUNY Press, 2002.

Donehower, Kim, Charlotte Hogg, and Eileen Schell. *Rural Literacies*. Southern Illinois UP, 2007.

Driskill, Qwo-Li. "Decolonial Skillshares: Indigenous Rhetorics as Radical Practice." *Survivance, Sovereignty, and Story: Teaching American Indian Rhetorics*, edited by Lisa King, Rose Gubele, and Joyce Rain Anderson, Utah State UP, 2015, pp. 57–78.

Dubisar, Abby M. "Cultivating Legitimacy as a Farmer." *Community Literacy Journal*, vol. 15, no. 2, 2021, pp. 31–51.

Eckstein, Justin, and Donovan Conley. "Spatial Affects and Rhetorical Relations at the Cherry Creek Farmers' Market." *The Rhetoric of Food: Discourse, Materiality, and Power*, edited by Joshua J. Frye and Michael S. Bruner, Routledge, 2012, pp. 171–89.

Edbauer, Jenny. "Unframing Models of Public Distribution: From Rhetorical Situation to Rhetorical Ecologies." *Rhetoric Society Quarterly*, vol. 35, no. 4, 2005, pp. 5–23.

Epps-Robertson, Candace. *Resisting Brown: Race, Literacy, and Citizenship in the Heart of Virginia*. U of Pittsburgh P, 2018.

Estabrook, Barry. *Tomatoland: How Modern Industrial Agriculture Destroyed Our Most Alluring Fruit*. Andrews McMeel Publishing, 2011.

Farmer, Frank. *After the Public Turn: Composition, Counterpublics, and the Citizen Bricoleur*. Utah State UP, 2013.

Feigenbaum, Paul. *Collaborative Imagination: Earning Activism through Literacy Education*. Southern Illinois UP, 2015.

Feigenbaum, Paul. "Tactics and Strategies of Relationship-Based Practice: Reassessing the Institutionalization of Community Literacy." *Community Literacy Journal*, vol. 5, no. 2, 2011, pp. 47–65.

Ferrante, Dana. "Disabled Activists Are Building a More Inclusive Food Justice Future." *Civil Eats*, 9 Nov. 2022, https://civileats.com/2022/11/09/disabled-activists-allies-building-inclusive-food-justice-systems-gardening-pantries-food-assistance/. Accessed 2 Dec. 2022.

Fishman, Charles. *The Big Thirst: The Secret life and Turbulent Future of Water*. Free Press, 2012.

Fishman, Jenn, and Lauren Rosenberg. "Community Writing, Community Listening." *Community Literacy Journal*, vol. 13, no. 1, 2018, pp. 1–6.

Flower, Linda. *Community Literacy and the Rhetoric of Public Engagement*. Southern Illinois UP, 2008.

Foley, Jonathan. "A Five Step Plan to Feed the World." *National Geographic*, 2015, https://www.nationalgeographic.com/foodfeatures/feeding-9-billion/.

Food Democracy Now! "Re: House Passes Dark Act 275 to 150 to Kill GMO Labeling." Email to Veronica House, 24 July 2015.

Food, Inc. Written and produced by Robert Kenner, Magnolia Pictures. 2008. DVD.

"Fort Chambers / Poor Farm Management Plan," https://bouldercolorado.gov/projects/fort-chambers-poor-farm-management-plan. Accessed 21 Oct. 2024.

Francis, Kerry, and Tim Francis. Interview. By Veronica House, 6 July 2020.

Frontline Farming. "Mission." https://www.frontlinefarming.org/mission. Accessed 12 Sept. 2024.

Frye, Joshua J., and Michael S. Bruner, editors. *The Rhetoric of Food: Discourse, Materiality, and Power*. Routledge, 2012.

García, Romeo. "Creating Presence from Absence and Sound from Silence." *Community Literacy Journal*, vol. 13, no. 1, 2019, pp. 7–15, https://doi.org/10.25148/clj.13.1.009086.

García, Romeo, and Damián Baca, editors. *Rhetorics Elsewhere and Otherwise: Contested Modernities, Decolonial Visions*. CCCC Studies in Writing and Rhetoric. National Council of Teachers of English, 2019.

Garth, Hanna, and Ashanté M. Reese, editors. *Black Food Matters*. U of Minnesota P, 2020.

Gayeton, Douglas. *Local: The New Face of Food and Farming in America*. Harper Design, 2014.

generative somatics. Website. https://generativesomatics.org.

Genoways, Ted. "Gagged by Big Ag." *Mother Jones*, Aug. 2013, p. 44+.

Gere, Anne Ruggles. "Kitchen Tables and Rented Rooms: The Extracurriculum of Composition." *CCC*, vol. 45, no. 1, 1994, pp. 75–92.

Gladwell, Malcolm. *The Tipping Point: How Little Things Can Make a Big Difference*. Little, Brown and Company, 2000.

Glennon, Robert. *Unquenchable: America's Water Crisis and What to Do About It*. Island Press, 2010.

Goldblatt, Eli. *Because We Live Here: Sponsoring Literacy beyond the College Curriculum*. Hampton Press, 2007.

Goldthwaite, Melissa A. *Food, Feminisms, Rhetorics*. Southern Illinois UP, 2017.

Gordon, Constance, Phaedra C. Pezzullo, and Michelle Gabrieloff-Parish, "Food Justice Advocacy Tours: Remapping Rooted, Regenerative Relationships Through Denver's 'Planting Just Seeds.'" *The Rhetoric of Social Movements: Networks, Power, and New Media*, edited by Nathan Crick, Routledge, 2021, pp. 299–316.

Gorsevski, Ellen W. "Chipotle Mexican Grill's Meatwashing Propaganda: Corporate-Speak Hiding Suffering of 'Commodity' Animals." *The Political Language of Food*, edited by Samuel Boerboom, Lexington Books, 2015, pp. 201–25.

Gottlieb, Robert, and Anupama Joshi. *Food Justice*. MIT Press, 2010.

Grabill, Jeffery T. *Writing Community Change*. Hampton Press, 2007.

Grace, Stephen. "The Hub of the Revolution." *Local Food Shift Colorado*, vol. 1, 2015, pp. 59–73.

Green, Keisha. "Doing Double Dutch Methodology: Playing with the Practice of Participant Observer." *Humanizing Research: Decolonizing Qualitative Inquiry with Youth and Communities*, edited by Django Paris and Maisha T. Winn, Sage Publishing, 2014, pp. 147–60.

Gries, Laurie E. "Agential Matters: Tumbleweed, Women-Pens, Citizen-Hope, and Rhetorical Actancy." *Ecology, Writing Theory, and New Media*, edited by Sidney I. Dobrin, Routledge, 2012, pp. 67–91.

Gries, Laurie E. *Still Life with Rhetoric: A New Materialist Approach for Visual Rhetorics*. Utah State UP, 2015.

Guthman, Julie. "'If They Only Knew': The Unbearable Whiteness of Alternative Food." *Cultivating Food Justice: Race, Class, and Sustainability*, edited by Alison Hope Alkon and Julian Agyeman, MIT Press, 2011, pp. 263–82.

Hahn, Laura K., and Michael S. Bruner. "Politics on Your Plate: Building and Burning Bridges Across Organic, Vegetarian, and Vegan Discourse." *The Rhetoric of Food: Discourse, Materiality, and Power*, edited by Joshua J. Frye and Michael S. Bruner, Routledge, 2012, pp. 42–57.

Hass, Angela M. "Wampum as Hypertext: An American Indian Intellectual Tradition of Multimedia Theory and Practice." *Studies in American Indian Literatures*, vol. 19, no. 4, Winter 2007, pp. 77–100. https://doi.org/10.1353/ail.2008.0005.

Hawhee, Debra. *A Sense of Urgency*. U of Chicago P, 2023.

Hawisher, Gail E., Cynthia L. Selfe, Brittney Moraski, and Melissa Pearson. "Becoming Literate in the Information Age: Cultural Ecologies and the Literacies of Technology." *College Composition and Communication*, vol. 55, no. 4, 2004, pp. 642–92.

Hayhoe, Katharine, and Andrew Farley. *A Climate for Change: Global Warming Facts for Faith-Based Decisions*. FaithWords, 2011.

Heinberg, Richard. "What Will We Eat as the Oil Runs Out?" The Lady Eve Balfour Lecture, 22 Nov. 2007. Web. 13 Dec. 2025. https://richardheinberg.com/188-what-will-we-eat-as-the-oil-runs-out.

Hesterman, Oran B., and Daniel Horan. "The Demand for 'Local' Food Is Growing—Here's Why Investors Should Pay Attention." *Business Insider*, 25 Apr. 2017, https://www.businessinsider.com/the-demand-for-local-food-is-growing-2017-4.

Holt-Giménez, Eric. "Food Security, Food Justice, or Food Sovereignty? Crises, Food Movements, and Regime Change." *Cultivating Food Justice: Race, Class, and Sustainability*, edited by Alison Hope Alkon and Julian Agyeman, MIT P, 2011, pp. 309–30.

hooks, bell. *Teaching Community: A Pedagogy of Hope*. Routledge, 2003.

Hoppe, Robert A. *Structure and Finances of US Farms: Family Farm Report*. USDA, Economic Research Service, 2014.

House, Veronica. "Community Engagement in Writing Program Design and Administration." *WPA: Writing Program Administration*, vol. 39, no. 1, 2015, pp. 54–71.

House, Veronica. "Keep Writing Weird: A Call for Eco-Administration and Engaged Writing Programs." *Community Literacy Journal*, vol. 11, no. 1, 2016, pp. 54–63.

House, Veronica. "The Reflective Course Model: Changing the Rules for Reflection in Service-Learning Composition Courses." *Reflections*, vol. 12, no. 2, 2013, pp. 27–65.

House, Veronica. "Re-Framing the Argument: Critical Service-Learning and Community-Centered Food Literacy." *Community Literacy Journal*, vol. 8, no. 2, 2014, pp. 1–16.

House, Veronica, with Sweta Baniya, Paul Feigenbaum, Megan Hartline, Lisa King, Seán McCarthy, Maria Novotny, Jessica Restaino, Sherita V. Roundtree, Daniel Singer, Lara Smith-Sitton, Karen Tellez-Trujillo, Don Unger, Bernardita M. Yunis Varas, Kate Vieira, Ada Vilageliu-Díaz, Stephanie Wade, and Christopher Wilkey. "Distributed Definition Building and the Coalition for Community Writing." *Peitho*, vol. 25, no. 4 (2023), https://cfshrc.org/article/distributeddefinition-building-and-the-coalition-for-community-writing/.

House, Veronica, and Stephanie Briggs. "Contemplative Practice for Community Building and Healing in Writing Studies: A Conversation." *Composition Forum*, 54 (Summer 2024): https://compositionforum.com/issue/54/contemplative-practice.php.

House, Veronica, and Kelly Zepelin. "When the Land Writes: The Rhetorical and Reciprocal Lives of Soil and Plants." *Food Justice Activism and Pedagogies: Literacies and Rhetorics for Transforming Food Systems in Local and Transnational Contexts*, edited by Eileen E. Schell, Dianna Winslow, and Pritisha Shrestha, Lexington Press, 2023, pp. 149–65.

Hsu, V. Jo. *Constellating Home: Trans and Queer Asian American Rhetorics*. The Ohio State UP, 2022.

Hubrig, Ada, and Christina Cedillo. "Access as Community Literacy: A Call for Intersectionality, Reciprocity, and Collective Responsibility." *Community Literacy Journal*, vol. 17, no. 1, 2022, pp. 1–8.

Hubrig, Adam. "'We Move Together': Reckoning with Disability Justice in Community Literacy Studies." *Community Literacy Journal*, vol. 14, no. 2, 2020, pp. 144–53, https://doi.org/10.25148/14.2.009042.

Indigenous Corporate Training. "What Is the Seventh Generation Principle?" May 30, 2020. https://www.ictinc.ca/blog/seventh-generation-principle.

Itchuaqiyaq, Cana Uluak. "No, I Won't Introduce You to My Mama: Boundary Spanners, Access, and Accountability to Indigenous Communities." *Community Literacy Journal*, vol. 17, no. 1, 2022, pp. 96–98.

Jackson, Rachel C., and Dorothy Whitehorse DeLaune. "Decolonizing Community Writing with Community Listening: Story, Transrhetorical Resistance, and Indigenous Cultural Literacy Activism." *Community Literacy Journal*, vol. 13, no. 1, 2018, pp. 37–54.

Jackson, Wes. "The 50-Year Farm Bill." *Solutions*, vol. 1, no. 3, 2010, pp. 28–35, www.thesolutionsjournal.com.

Jackson, Wes. "The Necessity and Possibility of an Agriculture Where Nature Is the Measure." *Conservation Biology*, vol. 22, no. 6, pp. 1376–77, https://doi.org/10.1111/j.1523-1739.2008.01101.x.

Jackson, Wes. "Tackling the Oldest Environmental Problem: Agriculture and Its Impact on Soil." *The Post Carbon Institute Reader Series*. Santa Rosa, 2010, pp. 1–9.

Kannan, Vani, Ben Kuebrich, and Yanira Rodriguez. "Unmasking Corporate-Military Infrastructure: Four Theses." *Community Literacy Journal*, vol. 11, no. 1, 2016, pp. 76–93, https://doi.org/10.25148/clj.11.1.009251.

Katz, Ruth. "You Are What You Environmentally, Politically, Socially, and Economically Eat: Delivering the Sustainable Farm and Food Message." *Environmental Communication*, vol. 4, no. 3, 2010, pp. 371–77. Web. https://doi.org/10.1080/17524032.2010.500451.

Kawagley, Angayuqaq Oscar. "Nurturing Native Languages." *Sharing Our Pathways*, vol. 7, no. 1, 2002, pp. 1–5. Web. https://www.uaf.edu/ankn/publications/collective-works-of-angay/Nurturing-Native-Languages.pdf.

Kelley, Robin D. G. *Freedom Dreams: Black Radical Imagination*. Beacon Press, 2003.

Kiehl, Jeffery T. *Facing Climate Change: An Integrated Path to the Future*. Columbia UP, 2016.

Kimmerer, Robin Wall. *Braiding Sweetgrass: Indigenous Wisdom, Scientific Knowledge and the Teachings of Plants*. Milkweed Editions, 2013.

Kimmerer, Robin Wall. "The Serviceberry: An Economy of Abundance." *Emergence Magazine*, 26 Oct. 2022. Web. https://emergencemagazine.org/essay/the-serviceberry/.

Kimmerer, Robin Wall. "We've Forgotten How to Listen to Plants." Interview. By Anne Strainchamps and Steve Paulson, July 2019. Web. https://www.ttbook.org/interview/weve-forgotten-how-listen-plants.

King, Lisa. "Sovereignty, Rhetorical Sovereignty, and Representation." *Survivance, Sovereignty, and Story: Teaching American Indian Rhetorics*, edited by Lisa King, Rose Gubele, and Joyce Rain Anderson, Utah State UP, 2015, pp. 17–34.

King, Lisa, Rose Gubele, and Joyce Rain Anderson. *Survivance, Sovereignty, and Story: Teaching American Indian Rhetorics*. Utah State UP, 2015.

King, Matthew Wilburn. "How Brain Biases Prevent Climate Action." *BBC*, 7 Mar. 2019, https://www.bbc.com/future/article/20190304-human-evolution-means-we-can-tackle-climate-change.

Kingsolver, Barbara. *Animal, Vegetable, Miracle: A Year of Food Life*. Harper Collins, 2007.

Kinloch, Valerie. *Harlem on Our Minds: Place, Race, and the Literacies of Urban Youth*. Teachers College Press, 2010.

Koch, Megan A., and Cristin A. Compton. "Corporate Colonization in the Market: Discursive Closures and the Greenwashing of Food Discourse." *The Political Language of Food*, edited by Samuel Boerboom, Lexington Books, 2015, pp. 227–50.

Kummer, Corby. "The Great Grocery Smackdown." *The Atlantic*, Mar. 2010, pp. 38–41.

Kynard, Carmen. "'All I Need Is One Mic': A Black Feminist Community Meditation on TheWork, the Job, and the Hustle (and Why So Many of Yall Confuse This Stuff)." *Community Literacy Journal*, vol. 14, no. 2, 2020, pp. 5–24, https://doi.org/10.25148/14.2.009033.

Kynard, Carmen. "'Oh No She Did NOT Bring Her Ass Up in Here with That!' Racial Memory Radical Reparative Justice, and Black Feminist Pedagogical

Futures." *College English*, vol. 85, no. 4, 2023, pp. 318–45. https://doi.org/10.58680/ce202332458.

Kynard, Carmen. "Teaching While Black." *Literacy in Composition Studies* 3, no. 1, 2015, pp. 1–20.

Kynard, Carmen. *Vernacular Insurrections: Race, Black Protest, and the New Century in Composition-Literacies Studies*. SUNY, 2013.

LaDuke, Winona. *All Our Relations: Native Struggles for Land and Life*. South End, 1999.

LaDuke, Winona. *Recovering the Sacred*. Haymarket Books, 2005.

Lakoff, George. *Don't Think of an Elephant! Know Your Values and Frame the Debate*. Chelsea Green, 2014.

Landrigan, Philip J., and Charles Benbrook. "GMOs, Herbicides, and Public Health." *New England Journal of Medicine*, vol. 373, no. 8, 20 Aug. 2015, pp. 263–65, https://www.nejm.org/doi/full/10.1056/NEJMp1505660.

Latour, Bruno. *Re-Assembling the Social*. Oxford UP, 2005.

Lertzman, Renee. *Environmental Melancholia: Psychological Dimensions of Engagement*. Routledge, 2015.

Levkoe, Charles Z., Peter Andree, Vikram Bhatt, Abra Brynne, Karen M. Davison, Cathleen Kneen, and Erin Nelson. "Collaboration for Transformation: Community-Campus Engagement for Just and Sustainable Food Systems." *Journal of Higher Education Outreach and Engagement*, pp. 32–61. Web. https://www.researchgate.net/publication/309731417_Collaboration_for_Transformation_Community-Campus_Engagement_for_Just_and_Sustainable_Food_Systems.

Lewis, Alan. "Coming Full Circle: Reclaiming Democracy with Local Food." *Local Food Shift Colorado*, 1, 2015, pp. 38–47.

Lien, Marianne E., and Gisli Pálsson. "Ethnography Beyond the Human: The 'Other-than-Human' in Ethnographic Work." *Ethnos*, vol. 86, no. 1, 2019, pp. 1–20, https://doi.org/10.1080/00141844.2019.1628796.

Loberg, Lindsey. Interview. By Veronica House, 6 June 2020.

Lockett, Alexandria L., Iris D. Ruiz, James Chase Sanchez, and Christopher Carter. *Race, Rhetoric, and Research Methods*. WAC, UP of Colorado, 2021.

Long, Elenore. *Community Literacy and the Rhetoric of Local Publics*. Parlor Press, 2008.

Low, Sarah A., Aaron Adalja, Elizabeth Beaulieu, Nigel Key, Steve Martinez, Alex Melton, Agnes Perez, Katherine Ralston, Hayden Stewart, and Shellye Suttles. *Trends in US Local and Regional Food Systems: Report to Congress*. USDA, Economic Research Service, 2015.

Magdoff, Fred, John Bellamy Foster, and Frederick H. Buttel. *Hungry for Profit: The Agribusiness Threat to Farmers, Food, and the Environment*. NYU P, 2000.

Magee, Rhonda. *The Inner Work of Racial Justice: Healing Ourselves and Transforming Our Communities Through Mindfulness*. TarcherPerigee, 2019.

Maiser, Jennifer. *Eat Local Challenge.* http://www.pbs.org/now/shows/344/locavore.html. Accessed 4 Mar. 2017.

Maraj, Louis M. *Black or Right: Anti/Racist Campus Rhetorics.* Utah State UP, 2020.

Marshall, George. *Don't Even Think About It: Why Our Brains Are Wired to Ignore Climate Change.* Bloomsbury, 2014.

Martinez, Aja. *Counterstory: The Rhetoric and Writing of Critical Race Theory.* CCCC/NCTE, 2020.

Martinez, Stephen, Michael S. Hand, Michelle Da Pra, Susan Pollack, Katherine Ralston, Travis Smith, Stephen Vogel, Shellye Clark, Luanne Lohr, Sarah A. Low, and Constance Newman. *Local Food Systems: Concepts, Impacts, and Issues.* United States Department of Agriculture. Economic Research Report, no. 97, 2010, https://www.ers.usda.gov/webdocs/publications/46393/7054_err97_1_.pdf?v=8070.2.

Mastrangelo, Lisa. "Community Cookbooks: Sponsors of Literacy and Community Identity." *Community Literacy Journal*, vol. 10, no. 1, 2015, pp. 73–86.

Mathieu, Paula. *Tactics of Hope.* Boynton/Cook Publishers, 2005.

Mathieu, Paula, and Angela Muir. "Introduction to the Special Issue: An Invitation to Contemplative Writing Pedagogy." *Composition Forum*, 54 (Summer 2024), https://www.compositionforum.com/issue/54/from-the-editors.php.

Mathieu, Paula, Steve Parks, and Tiffany Rousculp, editors. *Circulating Communities: The Tactics and Strategies of Community Publishing.* Lexington Books, 2012.

The Matrix. Produced by Joel Silver, Warner Bros., 1999.

McCrostie Little, Heather, and Nick Taylor. "Issues of New Zealand Farm Succession: A Study of the Integrational Transfer of the Farm Business." *MAF Policy Technical Paper.* Ministry of Agriculture, 1998.

McCullen, Christie. "The White Farm Imaginary: How One Farmers Market Refetishizes the Production of Food and Limits Food Politics." *Food as Communication/Communication as Food*, edited by Janet M. Cramer, Carlnita P. Green, and Lynn M. Walters, Peter Lang, 2011, pp. 218–34.

McDonald's. Website. "Frank Martinez," mcdonalds.com, https://www.mcdonalds.com/us/en-us/about-our-food/meet-our-suppliers/potatoes-frank-martinez.html.

McDonald's. Website. "McDonald's Food Suppliers," mcdonalds.com, https://www.mcdonalds.com/us/en-us/about-our-food/meet-our-suppliers.html.

McEntee, Jesse C. "Realizing Rural Food Justice: Divergent Locals in the Northeastern United States." *Cultivating Food Justice: Race, Class, and Sustainability*, edited by Alison Hope Alkon and Julian Agyeman, MIT P, 2011, pp. 239–59.

Meadows, Donella H. *Thinking in Systems.* Chelsea Green, 2008.

Medina, Johnitta. Email to author, 24 June 2024.

Menakem, Resmaa. *My Grandmother's Hands: Racialized Trauma and the Pathway to Mending Our Hearts and Bodies*, audiobook ed. Central Recovery Press, 2017.

Miele, Mara, and Adrian Evans. "When Foods Become Animals: Ruminations on Ethics and Responsibility in Care-*full* Practices of Consumption." *Ethics, Place and Environment*, vol. 13, no. 2, 2010, pp. 171–90, https://doi.org/10.1080/13668791003778842.

Mihesuah, Devon A., and Elizabeth Hoover. *Indigenous Food Sovereignty in the United States*. U of Oklahoma P, 2019.

Mitchell, Tania D. "Traditional vs. Critical Service-Learning: Engaging the Literature to Differentiate Two Models." *Michigan Journal of Community Service Learning*, vol. 14, no. 2, 2008, pp. 50–65.

Monberg, Terese Guinsatao. "Writing Home or Writing *as* the Community: Toward a Theory of Recursive Spatial Movement for Students of Color in Service-Learning Courses." *Reflections*, vol. 8, no. 3, 2009, pp. 21–51.

Moss, Beverly. *A Community Text Arises: A Literate Text and a Literacy Tradition in African-American Churches*. Hampton Press, 2003.

Moss, Beverly. *Literacy across Communities*. Hampton Press, 1994.

Nall, Madeline. Interview. By Veronica House. 8 July 2020.

National Young Farmers Coalition. 2017. "Land Access." youngfarmers.org, http://www.youngfarmers.org/policy/landaccess/. Accessed 14 Apr. 2018.

Nelsen, Kayla. "Indigenous Food Markets in Michigan Grow as Native Americans Reclaim Heritage." *Spartan News*, 5 Nov. 2021. Web. https://news.jrn.msu.edu/2021/11/indigenous-food-markets-in-michigan-grow-as-native-americans-reclaim-heritage/.

Nestle, Marion. "The Local Food Movement 15 Years In: Where Are We Now?" edible communities. Web. 6 Mar. 2017. https://ediblecommunities.com/edible-communities/local-food-movement-15-years/.

Nowacek, David M., and Rebecca S. Nowacek. "The Organic Foods System: Its Discursive Achievements and Prospects." *College English*, vol. 70, no. 4, 2008, pp. 403–19.

Opel, Dawn, and Donnie Sackey, editors. *Community Literacy Journal*, vol. 14, no. 1, 2019, pp. 1–6.

Ore, Ersula J. *Lynching: Violence, Rhetoric, and American Identity*. U of Mississippi P, 2019.

Ore, Ersula J. "Twenty-First Century Discourses of American Lynching." *Critical Discourse Studies*, 2022, https://doi.org/10.1080/17405904.2022.2090978.

Ore, Ersula, Kim Wieser, and Christina Cedillo. "Symposium: Diversity Is Not Enough: Mentorship and Community-Building as Antiracist Praxis." *Rhetoric Review*, vol. 40, 2021, pp. 207–56, https//doi.org/10.1080/07350198.2021.1935157.

Organic Produce Network. "State of Organic Produce 2023." https://eu-assets.contentstack.com/v3/assets/blt17bf506a5fa8d55b/blt32a5d17baaf8d2aa/6664bef4254a26a9435b8f0d/2023-state-of-organic-opn.pdf. Accessed 18 Dec. 2024.

Organic Trade Association. "Industry Report." 2018. Web. https://www.ota.com/resources/market-analysis. Accessed 21 Dec. 2018.

Owens, Derek. *Composition and Sustainability: Teaching for a Threatened Generation*. NCTE, 2001.

Pacheco-Borden, Carmen. Interview. By Veronica House, 10 Aug. 2020.

Paris, Django, Maisha T. Winn, and Keisha Green. "Doing Double Dutch Methodology: Playing with the Practice of Participant Observer." *Humanizing Research: Decolonizing Qualitative Inquiry with Youth and Communities*, Sage, 2018, pp. 147–60, https://doi.org/10.4135/9781544329611.n8.

Parks, Stephen. *Gravyland: Writing Beyond the Curriculum in the City of Brotherly Love*. Syracuse UP, 2010.

Parks, Steve, and Eli Goldblatt. "Writing Beyond the Curriculum: Fostering New Collaborations in Literacy." *Writing and Community Engagement: A Critical Sourcebook*, edited by Thomas Deans, Barbara Roswell, and Adrian J. Wurr, Bedford/St. Martin's, 2010, pp. 337–58.

Patel, Raj. *Stuffed and Starved*. Melville House, 2012.

Pauszek, Jessica. "Writing from 'The Wrong Class': Archiving Labor in the Context of Precarity." *Community Literacy Journal*, vol. 13, no. 2, 2019, pp. 48–68, https://doi.org/10.25148/clj.13.2.009069.

Peck, Wayne Campbell, Linda Flower, and Lorraine Higgins. "Community Literacy." *College Composition and Communication*, vol. 46, no. 2, National Council of Teachers of English, 1995, pp. 199–222, https://doi.org/10.2307/358428.

Peña, Devon. 2017. "The Hummingbird and the Redcap." *Wildness: Relations of People and Place*, edited by Gavin Van Horn and John Hausdoerffer, U of Chicago P, pp. 91–98.

Penniman, Leah. *Farming While Black*. Chelsea Green, 2020.

Penniman, Leah. "Why Food Is a Social Justice Issue." *The Doctor's Farmacy with Mark Hyman, M.D.*, 11 Dec. 2019. Podcast.

Plass, Tim. Email to Veronica House, 6 July 2023.

Pollan, Michael. "An Animal's Place." *The New York Times Magazine*, 10 Nov. 2002.

Pollan, Michael. *The Botany of Desire: A Plant's-Eye View of the World*. Random House, 2002.

Pollan, Michael. "No Bar Code." *Mother Jones*, May/June 2006. Accessed 28 Feb. 2011.

Pollan, Michael. *The Omnivore's Dilemma: A Natural History of Four Meals*. Penguin Press, 2006.

Pollan, Michael. "Our Decrepit Food Factories." *The New York Times Magazine*, 16 Dec. 2007. Web. 5 Nov. 2015. https://www.nytimes.com/2007/12/16/magazine/16wwln-lede-t.html.

Pollan, Michael. "Power Steer." *The New York Times*, 31 Mar. 2002. Web. https://www.nytimes.com/2002/03/31/magazine/power-steer.html.

Pollan, Michael. "Unhappy Meals." *The New York Times*. 28 Jan 2007. https://www.nytimes.com/2007/01/28/magazine/28nutritionism.t.html. Accessed 22 Nov. 2011.

Pollan, Michael. "Why 'Natural' Doesn't Mean Anything Anymore." *The New York Times Magazine*, 28 Apr. 2015. Web. https://www.nytimes.com/2015/05/03/magazine/why-natural-doesnt-mean-anything-anymore.html.

Poore, Joseph, and Thomas Nemecek. "Reducing Food's Environmental Impacts through Producers and Consumers." *Science*, vol. 360, no. 6392, 2018, pp. 987–92. Web. https://doi.org/10.1126/science.aaq0216.

Powell, Katrina M., and Pamela Takayoshi. "Accepting Roles Created for Us: The Ethics of Reciprocity." *College Composition and Communication*, 2003, pp. 394–422.

Powell, Malea. 2004. "Down by the River, or How Susan La Flesche Picotte Can Teach Us About Alliance as a Practice of Survivance." *College English*, vol. 67, no. 1, pp. 38–60.

Prentice, Jessica. "The Birth of Locavore." 20 Nov. 2007, https://blog.oup.com/2007/11/prentice/.

Primm, Annelle. *Racial Trauma and Climate Change*. The Medical Society Consortium on Climate and Health, 13 Aug. 2021. Webinar.

Pritchard, Eric Darnell. *Fashioning Lives: Black Queers and the Politics of Literacy*. Southern Illinois UP, 2017.

Project 95. Ollin Farms. Website. https://ollin-farms-project-95.square.site.

Rao, Vidya. "Black and Vegan: Why So Many Black Americans Are Embracing the Plant-Based Life." *Today*, 26 Feb. 2021, https://www.today.com/food/black-vegan-why-so-many-black-americans-are-embracing-plant-t209743.

Reiley, Laura. "Farm to Fable: At Tampa Bay Farm-to-Table Restaurants, You're Being Fed Fiction." *Tampa Bay Times*, 13 Apr. 2016, http://www.tampabay.com/projects/2016/food/farm-to-fable/restaurants/.

Rice, Jenny. *Distant Publics: Development Rhetoric and the Subject of Crisis*. U of Pittsburgh P, 2012.

Richardson, Elaine. "Coming from the Heart: African American Students, Literacy Stories, and Rhetorical Education." In *African American Rhetoric(s): Interdisciplinary Perspectives*, edited by E. B. Richardson and R. L. Jackson II, Southern Illinois UP, 2004, pp. 155–69.

Richardson, Elaine. "My *Ill* Literacy Narrative: Growing Up Black, Po and a Girl in the Hood." *Gender and Education*, vol. 21, no. 6, Nov. 2009, pp. 753–67.

Richardson, Elaine. "'She Ugly': Black Girls, Women in Hiphop and Activism—Hiphop Feminist Literacies Perspectives." *Community Literacy Journal*, vol. 16, no. 1, art. 3, 2021, https://digitalcommons.fiu.edu/communityliteracy/vol16/iss1/3.

Richardson, Elaine B., and R. L. Jackson II. *African American Rhetoric(s): Interdisciplinary Perspectives*, edited by E. B. Richardson and R. L. Jackson II, Southern Illinois UP, 2004.

Richardson, John H. "When the End of Human Civilization Is Your Day Job." *Esquire*, 20 July 2018. Web. 6 Aug. 2018. https://www.esquire.com/news-politics/a36228/ballad-of-the-sad-climatologists-0815/.

Ridolfo, Jim, and Danielle Nicole DeVoss. "Composing for Recomposition: Rhetorical Velocity and Delivery." *Kairos*, vol. 13, no. 2, 2009, n.p.

Riedel, Charlie. "Big Food's Biggest Trend? Campaigning Against Big Food." *Chicago Tribune*, Associated Press, 12 Sept. 2016.

Riley Mukavetz, Andrea, and Malea D. Powell. "Becoming Relations: Braiding an Indigenous Manifesto." *Decolonial Conversations in Posthuman and New Material Rhetorics*, edited by Jennifer Clary-Lemon and David M. Grant, The Ohio State UP, 2022, pp. 192–212.

Riley Mukavetz, Andrea, and Malea D. Powell. "Making Native Space for Graduate Students: A Story of Indigenous Rhetorical Practice." *Survivance, Sovereignty, and Story: Teaching American Indian Rhetorics*, edited by Lisa King, Rose Gubele, and Joyce Rain Anderson, Utah State UP, 2015, pp. 138–59.

Riley Mukavetz, Andrea, and Cindy Tekobbe. "'If You Don't Want Us There, You Don't Get Us': A Statement on Indigenous Visibility and Reconciliation." *Present Tense*, vol. 2, no. 9, 2022. Web. https://www.presenttensejournal.org/volume-9/if-you-dont-want-us-there-you-dont-get-us-a-statement-on-indigenous-visibility-and-reconciliation/.

Ríos, Gabriela. 2015. "Cultivating Land-Based Literacies and Rhetorics." *Literacy in Composition Studies*, vol. 3, no. 1, 2015, pp. 60–70, https://doi.org/10.21623/1.3.1.4.

Robin, Marie-Monique. *The World According to Monsanto*. 17 Feb. 2008. Film.

Rodgers, Ross. Interview. By Veronica House, 12 June 2020.

Rollins, Brooke, and Lee Bauknight, editors. *Food*. Fountainhead, 2010.

Roossien, Frances "Geri," and Andrea Riley Mukavetz. *You Better Go See Geri*. Oregon State UP, 2021.

Rose, Shirley K, and Irwin Weiser, editors. "Introduction: The WPA as Citizen-Educator." *Going Public: What Writing Programs Learn from Engagement*. Utah State UP, 2010, pp. 1–14.

Rosiek, Jerry Lee, Jimmy Snyder, and Scott L. Pratt. "The New Materialisms and Indigenous Theories of Non-Human Agency: Making the Case for Respectful Anti-Colonial Engagement." *Qualitative Inquiry*, vol. 26, nos. 3–4, 2019, pp. 331–46. Web. https://doi.org/10.1177/1077800419830135.

Rosset, Peter M., and Miguel A. Altieri. *Agroecology: Science and Politics*. Fernwood Publishing, 2017.

Rousculp, Tiffany. *Rhetoric of Respect: Recognizing Change at a Community Writing Center*. CCCC/NCTE, 2014.

Royster, Jacqueline Jones. "Disciplinary Landscaping, or Contemporary Challenges in the History of Rhetoric." *Philosophy and Rhetoric*, vol. 36, no. 2, 2003, pp. 148–67.

Royster, Jacqueline Jones. *Traces of a Stream: Literacy and Social Change Among African American Women*. U of Pittsburgh P, 2000.

Ruiz, Iris D. *Reclaiming Composition for Chicano/as and Other Ethnic Minorities*. Palgrave Macmillan, 2016.

Ruiz, Iris, and Sonia Arellano. "*La Cultura Nos Cura*: Reclaiming Decolonial Epistemologies through Medicinal History and Quilting as Method." *Rhetorics Elsewhere and Otherwise*, edited by Damián Baca and Romeo García, National Council of Teachers of English, 2019, pp. 141–68.

Ruiz, Iris, and Damián Baca. "Decolonial Options and Writing Studies." *Composition Studies*, vol. 45, no. 2, 2017, pp. 226–29.

Ryder, Phyllis Mentzell. *Rhetorics for Community Action: Public Writing and Writing Publics*. Lexington Books, 2011.

Sackey, Donnie Johnson. *Trespassing Natures: Species Migration and the Right to Space*. The Ohio State UP, 2024.

Sackey, Donnie Johnson, and Danielle Nicole DeVoss. "Ecology, Ecologies, and Institutions: Eco and Composition." *Ecology, Writing Theory, and New Media*, edited by Sidney I. Dobrin, Routledge, 2012, pp. 195–211.

Salmón, Enrique. *Eating the Landscape: American Indian Stories of Food, Identity, and Resilience*. U of Arizona P, 2012.

Salmón, Enrique. "Kincentric Ecology: Indigenous Perceptions of the Human–Nature Relationship." *Ecological Applications*, vol. 10, no. 5, 2000, pp. 1327–32.

Salmón, Enrique. "No Word." *Wildness*, edited by Gavin Van Horn and John Hausdoerffer, U of Chicago P, 2017, pp. 24–32.

Schell, Eileen E. "The Rhetorics of the Farm Crisis: Toward Alternative Agrarian Literacies in a Globalized World." *Rural Literacies*, by Kim Donehower, Charlotte Hogg, and Eileen Schell. Southern Illinois UP, 2007, pp. 77–119.

Schell, Eileen, Dianna Winslow, and Pritisha Shrestha. *Food Justice Activism and Pedagogies*. Lexington Books, 2023.

Schlosser, Eric. *Fast Food Nation*. Mariner Books, 2001.

Seas, Kristin. "Writing Ecologies, Rhetorical Epidemics." *Ecology, Writing Theory, and New Media*, edited by Sidney I. Dobrin, Routledge, 2012, pp. 51–66.

Shah, Rachael W. *Rewriting Partnerships: Community Perspectives on Community Learning*. Utah State UP, 2020.

The Shed. Website (site discontinued).

Sherman, Sean. *The Sioux Chef's Indigenous Kitchen*. U of Minnesota P, 2017.

Shiva, Vandana. "Biopiracy: The Colonization of the Seed." Schumacher Center Lecture, Mar. 2000, Great Barrington, MA.

Shiva, Vandana. "Diverse Expressions of a Living Earth." *for the wild (311)*, 2 Nov. 2022. Web. https://forthewild.world/listen/vandana-shiva-on-diverse-expressions-of-a-living-earth-311.

Shiva, Vandana. *Seed Sovereignty, Food Security*. North Atlantic Books, 2016.

Shiva, Vandana. *Staying Alive: Women, Ecology, and Development*. Zed Books, 1989.

Shiva, Vandana. *Water Wars: Privatization, Pollution, and Profit*. North Atlantic Books, 2002.

Shiva, Vandana. *Who Really Feeds the World?* North Atlantic Books, 2016.

Shultz, Kara. "On Establishing a More Authentic Relationship with Food: From Heidegger to Oprah on Slowing Down Fast Food." *The Rhetoric of Food: Discourse, Materiality, and Power*, edited by Joshua J. Frye and Michael S. Bruner, Routledge, 2012, pp. 223–37.

Simpson, Leanne Betasamosake. "Land as Pedagogy: Nishnaabeg Intelligence and Rebellious Transformation." *Decolonization: Indigeneity, Education and Society*, vol. 3, no. 3, 2014, pp. 1–25.

Smith, Linda Tuhiwai. *Decolonizing Methodologies: Research and Indigenous Peoples*, 2nd ed. Zed Books, 2012.

Smith, Mistinguette. "Wild Black Margins." *Wildness: Relations of People and Place*, edited by Gavin Van Horn and John Hausdoerffer, U of Chicago P, 2017, pp. 137–44.

Snider, Laura. "Boulder County Agrees to Allow Some GMOs on Public Land." *Daily Camera*, 20 Dec. 2011. Web. 7 Aug. 2015. https://www.dailycamera.com/ci_19585517/.

Stebbins, Samuel, and Michael B. Sauter. "25 Richest Cities in America: Does Your Metro Area Make the List?" *USA Today*. 23 May 2018. Web. https://www.usatoday.com/picture-gallery/money/2019/05/19/25-richest-cities-in-america/39483351/.

Stoknes, Per Espen. *What We Think About When We Try Not to Think About Global Warming*. Chelsea Green, 2015.

Stroh, David Peter. *Systems Thinking for Social Change*. Chelsea Green, 2015.

Stuckey, J. Elspeth. *The Violence of Literacy*. Heinemann, 1990.

Syverson, Margaret A. *The Wealth of Reality: An Ecology of Composition*. Carbondale: Southern Illinois UP, 1999.

Taczak, Kara, Veronica House, and Sheila Carter-Tod. "From Chaos Emerges Something New, Something Better—Something We Haven't Even Imagined." *There Are Writing Emergencies: Composing (Ourselves) in Times of Crisis*, edited by Holly Hassel and Kate Lisbeth Pantelides (forthcoming).

TallBear, Kim. "Why Interspecies Thinking Needs Indigenous Standpoints." *Society for Cultural Anthropology*, 18 Nov. 2011.

10 Million Black Vegan Women. Website. https://10millionblackveganwomen.org.

Thompson, John R. "'Food Talk': Bridging Power in a Globalizing World." *The Rhetoric of Food: Discourse, Materiality, and Power*, edited by Joshua J. Frye and Michael S. Bruner, Routledge, 2012, pp. 58–70.

"3 Nebraskans Plead Guilty in Organic Fraud Scheme." *Omaha World Herald*, 13 Oct. 2018, https://omaha.com/eedition/sunrise/articles/3nebraskans-pleadguilty-in-organic-fraudscheme/article_4efa6a86-24ae-5f40-839b-fa155e4ca3fe.html.

Todd, Zoe. "An Indigenous Feminist's Take on the Ontological Turn: 'Ontology' Is Just Another Word for Colonialism." *Journal of Historical Sociology* 29, no. 1, 2016, pp. 4–22.

Trauth, Erin. "Nutritional Noise: Community Literacies and the Movement Against Foods Labeled as 'Natural.'" *Community Literacy Journal*, vol. 10, no. 1, 2015, pp. 4–20.

UN Climate Change Annual Report, 2018. Web. https://unfccc.int/sites/default/files/resource/UN-Climate-Change-Annual-Report-2018.pdf.

UN Intergovernmental Panel on Climate Change. *Climate Change 2014: Impacts, Adaptation, and Vulnerability*. 5 Nov. 2015. Web. https://www.ipcc.ch/report/ar5/wg2/.

Urban Roots. Produced by Mathew Schmid, Leila Conners, and Mark McInnis, Tree Media, 2013. Documentary.

US Census of Agriculture. 2017. "Farm Demographics—US Farmers by Gender, Age, Race, Ethnicity, and More." United States Department of Agriculture, https://www.agcensus.usda.gov/Publications/2012/Online_Resources/Highlights/Farm_Demographics/. Accessed 27 Nov. 2017.

USDA. "Local Foods." *usda.gov*, 2017, https://www.nal.usda.gov/aglaw/local-foods#quicktabs-aglaw_pathfinder=1. Accessed 12 Dec. 2017.

USDA. *2012 Census of Agriculture: Colorado State and County Data*. Edited by USDA. 2014a. https://agcensus.library.cornell.edu/census_parts/2012-state-and-county-profiles-colorado/.

USDA Economic Research Service. "Disability Status Can Affect Food Security Among U.S. Households." usda.gov, 2022, https://www.ers.usda.gov/data-products/chart-gallery/gallery/chart-detail/?chartId=105136. Accessed 21 Nov. 2022.

Wade, Breeshia. *Grieving While Black: An Antiracist Take on Oppression and Sorrow*, audiobook ed. North Atlantic Books, 2021.

Wade, Stephanie. "Feed Your Mind: Cultivating Ecological Community Literacies with Permaculture." *Community Literacy Journal*, vol. 10, no. 1, 2015, pp. 87–98.

Walker, Kenneth. *Climate Politics on the Border: Environmental Justice Rhetorics*. U of Alabama P, 2022.

Walsh, Bryan. "The Real Cost of Cheap Food." *Time Magazine*, 31 Aug. 2009, pp. 30–37.

Washington, Karen. *FAQs*. https://www.karenthefarmer.com/faq-index.

Watts, Vanessa. "Indigenous Place-Thought and Agency Amongst Humans and Non Humans (First Woman and Sky Woman Go On a European World Tour!)." *Decolonization: Indigeneity, Education and Society*, vol. 2, no. 1, 2013, pp. 20–34.

Weisser, Christian R., and Sidney I. Dobrin, editors. *Ecocomposition*. SUNY, 2001.

White, Monica M. *Freedom Farmers: Agricultural Resistance and the Black Freedom Movement*. UNC P, 2018.

White, Rowen, guest. "Seeds, Grief, and Memory with Rowen White." *Finding Our Way Podcast*. Season 2, episode 6. 24 May 2021. https://www.findingourwaypodcast.com/individual-episodes/s2e6. Accessed 18 Dec. 2024.

Williamson, James. 2017. "Beginning Farmers and Age Distribution of Farmers." United States Department of Agriculture, last modified 13 Apr. 2017. https://www.ers.usda.gov/topics/farm-economy/beginning-disadvantaged-farmers/beginning-farmers-and-age-distribution-of-farmers/. Accessed 26 Nov. 2017.

Wolfenbarger, Deon. *Boulder County's Agricultural Heritage*. Report prepared for Boulder County Parks and Open Space and Boulder County Land Use Department, 10 Mar. 2006, https://assets.bouldercounty.org/wp-.

Wolynn, Mark. *It Didn't Start with You: How Inherited Family Trauma Shapes Who We Are and How to End the Cycle*. Penguin Life, 2017.

Yates, Phillip. Email to Veronica House. 7/30/2024.

Young, Anna "Amy." "Rhetorics of Smoke and Cedar: The Terroir of Texas BBQ." *Inventing Place: Writing Lone Star Rhetorics*, edited by Cayce Boyle and Jenny Rice, Southern Illinois UP, 2018, pp. 38–46.

Index

Page numbers followed by f indicate figures, page numbers followed by n indicate notes, and page numbers followed by t indicate tables.

Abisaid, Joseph L., 50, 57
ableism, 8, 18, 24, 41, 43, 57, 207, 243
abundance, 45, 206; critical, 111, 125, 127, 138, 139, 163, 166, 171, 175, 181, 184, 187, 213; rhetorics of, 27, 109, 111, 113, 123, 126, 130, 132–38, 139, 163, 166, 171, 175, 181, 184, 185, 187, 196, 213
Academic Time, 17
accountability, 9, 19, 21, 26, 34, 38, 109, 142, 173, 184, 208, 209, 210, 211, 216, 236; community, 237; desire for, 33; obligation for, 35; reconnecting with, 237; relational, 126–27
Acequia Institute, 211
Acosta, Angel, 121, 125
activism, 30, 144; food justice, 174; Indigenous, 130; local food, 10, 110, 112–13, 198; term, 211
affordability, 67, 68, 97, 99, 225
agency, 209, 210, 215; Anishinaabe understanding of, 221; authorial, 219; beyond-human creative, 220; deriving, 221; land-based/plant-based, 228; other-than-human, 219, 220; rhetorical, 231
agricultural heritage, 37, 79, 80, 208
agriculture, 36, 65, 109, 111, 119, 134–35, 157, 208; African, 35, 40; biodiversity and, 47; climate change and, 44; complexities of, 163; European American, 35, 40; food literacy and, 31; innovation in, 51, 52; local, 61, 78, 153, 166; patriarchal, 54; preserving, 78; racism and, 233; regenerative, 81, 165, 208, 216, 229; support for, 80–81; sustainable, 51, 52, 82–83; white settler colonialism and, 233. *See also* industrial agriculture; organic agriculture
Agriculture and Human Values, 52
Agyeman, Julian, 57, 66, 67, 68
Ahmed, Sarah, 211–12
Alexander, Michelle, 234
Alkon, Alison, 57, 66, 67, 68
All Our Relations (LaDuke), 139
American Home Products, 54
American Indian Resources, 212

271

American Indian Tribal Nations, 212
American Sign Language (ASL), 69
ancestral legacies, 38, 123, 130, 244, 245–46
Anderson, Joyce Rain, 150, 218, 220, 246n2
anger, 9, 32, 113, 122
animacy: grammar of, 222, 223, 224; Indigenous scholarship on, 29; other-than-human, 227; respect for, 221–22
animals, 56; co-writing by, 215; maltreated, 202
anthropomorphizing, 223, 224, 227
anti-ableism, 18, 24
antibiotics, 44, 48, 49
anticolonialism, 18, 130
antiracism, 16, 18, 66, 130, 232
anxiety, 96, 112, 119, 121, 136; climate, 117; environmental, 113
Arapaho, 35, 36, 37
Arellano, Sonia, 221, 228
art show, 190, 198f; handout for, 196f; local food and, 198
artwork, 215; definition building and, 188–91, 193–207
awareness, 4, 77, 93, 133, 216, 227, 232; consumer, 27; growth, 29, 246; public, 156; raising, 42, 125, 176, 177, 180; rhetorical, 193, 211
Aztecs, 242

Baca, Damián, 222
Bauer, Nate, 24
Bayer Group, 51, 52
behavior, 27, 114, 121, 127, 130, 147, 149, 160, 176; changing, 118–19, 210; ethnically questionable, 50; models of, 26; patterns of, 117; psychology and, 112; social, 125
Benjamin, Ruha, 234
Bennett, Jane, 146
Berger, Jonah, 176, 177
Bernardo, Shane, 169, 241
Berry, Wendell, 65
"Beyond Organic" (Coleman), 56
biases, 115, 119, 121
big agriculture, 40, 47, 53, 55, 61
Big Local, 61
big organic, 55, 58, 59

biodiversity, 37, 47, 56
BIOMIMICZ, 129
BIPOC advocates, 15, 37, 212, 234, 235, 246, 247
Black Food Matters (Garth and Reese), 174
Black Latinx Farmers Immersion, 71
Blackbelly Market, 204, 205
Blank Sanctuary Gardens, 68
blogs, writing for, 178–79
Bloom, Lynne, 66
bodies: Black/criminalization of, 234; land and, 223; nourishing, 228
Born, Branden, 62, 100
Boulder. *See* City of Boulder
Boulder County: agriculture in, 134–35, 153, 154; climate change and, 119; cultural variation of, 102–3; deficiency/abundance in, 132–38; ethos of, 209; farming in, 58, 156; food insecurity in, 135, 175; food literacy in, 5, 31–32, 112, 180; local food in, 4, 11, 19, 21, 23–24, 26, 27, 32, 38, 75–78, 91, 98, 100, 105, 108, 110, 145, 151, 162, 170, 171; mapping of, 168; Native Nations and, 37; Open Space and, 153; resilience/struggle in, 8; rhetoric/practice of, 78
Boulder County Commission, 134
Boulder County Comprehensive Plan, 78
Boulder County Farmers Markets, 62, 94, 97, 109, 131, 132, 151, 154, 156, 185
Boulder County Land Use Code, 165
Boulder County Open Space, 36, 77, 137, 146, 154, 155, 165, 212; agriculture on, 81, 153; GMOs and, 78, 79, 80, 134; land uses in, 78; maintenance of, 153; preservation of, 147, 215
Boulder County Parks and Open Space, 100, 165
"Boulder County Parks and Open Space Cropland Policy," 81
"Boulder County's Agricultural Heritage," 79
Boulder Daily Camera, 88
Boulder Farmers Market, 91, 100; gaps at, 96–97
Boulder Food Rescue, 41, 97, 180, 181, 213
Boulder Local Food Shift, 77

Boulder Museum of Contemporary Art, 193
Boulder Public Library, 189
Boulder Valley Comprehensive Plan, 153
Boulder Valley Public Schools, 188
Boulder Valley School District, 132
Boulder Weekly, 88
BoulderTalks, 164
Braiding Sweetgrass (Kimmerer), 34, 218, 222
Brandt, Deborah, 53, 170
Bridge House, 97, 172
Broad, Garrett, 210, 211
brochures, 194f, 195f
brown, adrienne maree, 9n3, 21, 33, 124–25, 136, 138, 139; on collective action, 216; on disasters/elite, 118; emergent strategy and, 157, 158
Brownlee, Michael, 77, 100, 109
Bruner, Michael S., 128
bumper stickers, messages on, 179–80
Business Insider, 60
"Buy Local" promotion, 61

CAFOs. *See* concentrated animal feeding operations
California Certified Organic, 55
canning, 180, 210
capitalism, 50, 51, 122, 242; corporate, 24, 59, 109, 135, 148; racial, 109
carbon footprint, 50, 94, 100, 199, 202, 203, 204
carbon sequestration, 48, 155
Carmen's Salsa, 101, 103, 105
Carnegie Library for Local History, 33
Carver, George Washington, 54
Castillo, Carl, 152
cattle, factory farming of, 148
Cavem, DJ, 72, 129
Cedillo, Christina, 18
Center for Communication and Democratic Engagement, 164
Center for Community Engagement, 213
Center for Food Safety, 48
Center for Story-Based Strategy, 124
Chef Ann Foundation, 188, 193
Cherokee Nation, 48
Cheyennes, 35, 36, 37

Chipotle, 50
cider press, 181, 183f
City of Boulder, 31, 35, 37, 76, 98, 164; American Indian Tribal Nations and, 212; definitional complexities and, 88; history of, 36, 77; land acquisition by, 36; local food in, 3, 98, 180; median income in, 96; Native Americans and, 211; poverty rate in, 96; Staff Land Acknowledgment by, 33; stewardship and, 211
City-Tribal Consultations, 212
Claiborne, Jenné, 72
climate, 216; co-writing by, 215
climate change, 8, 46, 51, 52, 70, 110, 111, 117, 118, 119, 180, 229, 233; agricultural yield and, 44; denying, 115; evidence of, 116; focus on, 47; food literacy and, 128–32; mental health consequences of, 113; people of color and, 120
climate communication, 111, 123, 128, 136, 139, 150, 176; focus of, 115; lessons of, 112–20; psychological responses to, 113–20; studies about, 108, 113
climate crisis, 110, 113, 119
climate scientists, 113, 114, 125, 135
co-authorship, 231
co-creation, 221, 223
co-shaping, 221
co-writing, 215
Coalition for Community Writing, 17n6
Coleman, Eliot, 56
Coleman-Jensen, Alisha, 43
collaboration, 9, 10, 13, 15, 16, 17, 21, 27, 73, 75, 106, 120, 124, 126, 127, 133, 136, 137, 139, 143, 151–56, 158, 164, 184; community and, 128; community writing and, 26; cultivating cultures of, 217; models of, 151; other-than-human, 19
Collaborative Imagination (Feigenbaum), 20
colonial paradigms, 207, 230, 243
colonialism, 228; legacy of, 40; ongoing, 41; reinforcing, 243; white settler, 34, 79, 242
colonization, 8, 36, 40, 51, 78, 141, 242; development of, 79; farm, 34; histories of, 37

Colorado State University Extension, 154, 165
Colorado Territory, 36
communication, 28, 134, 136, 160, 176, 177, 221, 227; building, 5; climate, 27, 114; controlling, 51; cycles of, 159; effective, 113, 116; environmental, 116, 152; strategies, 151; technical, 12
Communication and Environmental Studies, 13
community, 22, 116, 125, 140, 222–23; collaboration and, 128; embeddedness of, 23; land and, 232
community building, 28, 30, 38, 191, 210, 232; collaborative, 184, 216
community discourse analysis, 11; media and, 89
Community Food Share, 172
Community Fruit Rescue, 181
community literacy, 7, 14, 15, 32, 142, 143, 149–50, 151, 219; events, 210
Community Literacy Journal, 15, 219
Community Services Unlimited, 210
community supported agriculture (CSAs), 60, 78, 158, 178, 193, 225
Community Table Kitchen, 172
community writing, 4, 13, 29, 73, 83, 122, 125, 139, 140, 142, 159, 246; decolonizing, 240; distributed, 240; ecological approach to, 20–24; as evolving/relational work, 14–20; paradigms for, 220; practitioners, 107, 123, 126, 128, 140, 143, 145, 147, 157, 162, 215; theories/practices/priorities and, 7, 25
Composition Studies, social turn in, 140
composting, 54, 85
Compton, Cristin, 86
Conagra, 100
concentrated animal feeding operations (CAFOs), 44, 48, 50
Conference on College Composition and Communication, 24
Conference on Community Writing, 14, 24
Conley, Donovan, 91
connections, 9, 69, 116, 126, 150, 168, 237; building, 219; community, 129, 208; emotional, 4; human-to-human, 106, 140, 232; linguistic, 106; nature-based, 106; physical, 106
Constellating Home (Hsu), 107
consumers: ethically non-competent, 86; grassroots movement among, 57; organic, 54
contagion theory, 28, 159–61, 176
Contagious: Why Things Catch On (Berger), 176
Cooper, Ann, 188, 193
Cooper, Marilyn, 143–44, 145
Coppom, Brian, 62–63, 97, 109, 151, 152, 153; primal connection and, 131
corn, 48; animal feed and, 52; ethanol and, 52; genetically modified, 49, 52, 79, 80, 82
Corn Mother stories, 150
Costco, 100, 101, 103
cottage food laws, 90
County Sustainable Food and Agriculture Fund, 80
Covey, Stephen, 16
COVID-19, 114, 211, 243
critical race theory, 128
crops, 48, 54; drought-tolerant, 52; genetically modified, 81, 82, 134
cryptosporidium, 44
CSAs. *See* community supported agriculture
cultural issues, 6, 25, 36, 47
cultural systems, 116, 225
culture, 120, 125, 146, 221, 235; American, 106, 127; Indigenous, 6n2, 126, 127, 141, 168, 185, 230, 241; local food, 101; loss of, 19, 36; Mexican, 101; modern, 232; national, 63; network, 160; Potawatomi, 127; settler colonial, 33; white supremacy, 111, 118, 243; writing and, 145. *See also* food culture
Curanderismo, 242

D-Town Farms, 68, 130
Dakota Territory, 35
Dancer, Ashley, 152, 153, 155, 157, 163, 177; food literacy and, 159; interviews by, 154; thesis of, 156
Davis, Angela, 136, 139

Deans, Tom, 32, 145, 220
decolonization, 24, 131, 220, 221, 242
Deetz, Stanley, 51
definition building, 5, 21, 132, 158, 187, 209, 219; innovations/reclamations in, 68–73; method of, 5n1, 30; power of, 177; public art and, 188–91, 193–207
definitions: dynamic, 214–17; importance of, 50–53; multiple, 89
DeLaune, Dorothy Whitehorse, 240
Deleuze, Gilles, 140
Deloria, Vine, Jr., 219–20, 222
Denver Post, The, 88
Detroit Black Community Food Security Network, 68
Dharma's Garden, 185, 186f, 224, 225, 226f, 227, 238–39, 238f, 239f, 240; food stand of, 241f
diet: choices/preferences, 58, 107, 112; decolonization of, 24; Indigenous, 131
Dig In! Art Contest, 190
Dig In! project, 28, 152, 156, 158, 167, 172, 175, 185, 188, 190, 225; brochures/art for, 213; grants from, 179, 180, 182, 183, 189; launch of, 13
"Digging Deeper: A Food Justice Perspective" (Constance), 189
Dingo, Rebecca, 162–63
disabilities, 42, 69; food insecurity and, 43
disconnection, 5, 38, 64, 232–33, 243
discourse analysis, 88–91, 99
discourses, 11, 88–91, 99, 144; dominant, 98; local, 89, 97, 151; media, 159; national, 39–40; public, 25, 28, 112, 175; rhetorical, 112, 145
disease, 8, 48, 51, 71, 110; food-related, 5
dispossession, 19, 242, 243
distributed definition building, 4–5, 21, 24, 27, 28, 29, 42, 45, 63, 69, 109, 142, 152, 159, 173, 177, 188, 206, 213, 214, 237; accountability in, 209; breaking barriers to, 12; collective/collaborative ideation and, 137; critically framed/abundant approach to, 8–9; developing, 5, 180; ecological community writing and, 6–7, 9, 11, 76, 106, 210; example of, 93f; flow of, 163; focus on, 246; importance of, 76; justice/accountability/transformation and, 9; knowledge co-construction and, 10; local food and, 171; method for, 88–96, 113; participation in, 223, 229; projections by, 111; reciprocity and, 224–36; term, 98
distributed writing, 23, 175–81, 184–85, 187–91, 193–207
"Diverse Expressions of a Living Earth" (Shiva), 39
Dobrin, Sidney, 144, 145, 161
Don't Even Think About It: Why Our Brains Are Wired to Ignore Climate Change (Marshall), 114
Don't Think of an Elephant (Lakoff), 114
Double-Up Food Bucks, 80, 91, 97
Driskill, Qwo-Li, 221, 244

E. coli, 44, 49, 202
"Early Settlement/Pioneer Agriculture" (Wolfenberger), 79
Eat Local Challenge, 59, 87
Eat Local Project, 104
Eckstein, Justin, 91
ecocomposition, 141, 143, 144
ecological community writing, 14, 19, 24, 25, 30, 35, 39–40, 89, 110, 111, 120–23, 136, 138, 140, 151–52, 177, 188, 229, 246; abundance and, 171; ancestral trauma/grief and, 121; approach to, 159, 165; collaborative/interdisciplinary/intercommunity, 27; community and, 219; courses/projects, 174; critical, 175; developing, 76, 83, 150; distributed definition building and, 6–7, 9, 11, 76, 106, 210; enacting, 197; goals of, 148; importance of, 22–23; justice and, 10, 207–14; methodology of, 4–5, 21–22, 26–27, 28, 29, 43, 83, 106, 132, 142, 143, 161, 173, 212, 220, 223, 243, 244–45, 247; potential/energy of, 210; teaching, 73–74; thinking about, 159
Ecological Kitchen, The, 213
ecological systems, 144, 145, 147, 160
ecological theory, 28, 140, 141, 144, 162
ecology, 16, 19, 20, 28, 29–30, 42, 43, 128, 142, 143, 147, 149, 157, 165, 173;

abundant, 215; affective, 120–23, 140; assessing, 151–56; building, 163, 175–81, 184–185, 187–91, 193– 207, 220; coalitional, 4; complicated, 139, 162; connected, 89, 246; defining, 38; dynamic, 214–17; ecocomposition, 144; expansive/dynamic, 10; food, 75, 108; historical trauma and, 120–23; kin-centric, 230; linguistic, 140; local, 89, 140, 169; local food, 10, 83, 101, 106, 132, 136, 177, 179, 181, 191; mapping of, 168; rhetorical, 140, 145, 161; visualization of, 191; working for, 169. *See also* local ecology

economic systems, commoditization by, 128

ecopsychology, 7, 27, 111, 128, 134, 139, 149, 176, 184, 208, 210; lessons from, 112–20; studies about, 113. *See also* psychology; social psychology

ecosystems, 119, 128, 228; industrial agriculture and, 148; local, 95; terrestrial/aquatic, 47

Edbauer, Jenny, 145

education, 185; civic, 30; consumer-directed, 94; critical, 210; food literacy, 206; higher, 240, 243; local food, 89, 101, 151, 166; public, 99

Education and Promotion Committee, 133, 134, 190

Ellis, Bronwyn: work of, 200f

Emergency Family Assistance Association, 97

Emmad, Fatuma, 67

emotions, 4, 64, 107; complexity of, 123, 136; defensive, 118; harnessing, 166; intergenerational, 121; negative, 96

empathy, 126, 128, 137–38

engagement, 4, 11, 12, 32, 176, 178, 214; collaborative, 136; community, 7, 14, 15, 16, 17, 26, 75, 113, 157, 227; ecological, 22, 75; embodied, 184; encouraging, 136, 138; ethical, 25; pedagogical, 123; trans-community, 219; turning away from, 8

Environmental Center, 152

environmental degradation, 46, 47, 49, 59, 62, 64, 124, 148, 172, 225

environmental issues, 4, 12, 25, 26, 46, 55, 61, 94–95, 104, 107, 108, 109, 111, 112, 122, 123, 144, 149, 156, 157

Environmental Melancholia (Lertzman), 117

Environmental Studies Program, 152, 157, 163

Environmental Working Group, 57

ethics, 62, 63, 77–83, 101, 142, 198; financial means to, 108

Evans, Adrian, 86

Everett, Daniel, 127

exploitation, xii, 5, 8, 26, 53, 221, 228, 235, 243

"Exploring Solutions to Boulder County Agricultural Sector Constraints" (CSU Extension), 165

Facing Climate Change (Kiehl), 127

Farm Act (2008), 59–60

Farm Bill-Organic Food Productions Act (1990), 55

farmers market patrons: conversations with, 92–93; responses from, 93f, 94, 95t

farmers markets, 92f, 99, 101–2, 105, 111, 158, 159, 185, 193; literacy project for, 91–96

Farmers Market's Seeds Café, 189

farming: carbon, 155; factory, 48, 105, 148, 244; GMO, 82, 135; local, 3, 155; organic, 134, 137; regenerative, 81, 98, 224; vertical, 89

Farming While Black (Penniman), 71

farmworkers, 53, 103, 134, 190

Fast Food Nation (Schlosser), 100

feedlots, 48–49, 148

Feigenbaum, Paul, 20, 135, 214, 231

Ferrante, Dana, 43

fertilizers, 47, 100

First Nations Development Institute, 37, 211

Fisher, Walter, 124

Fishman, Jenn, 21

flooding, 35, 43, 46, 77, 83, 87

Flower, Linda, 23–24

Floyd, George, 233

Fobes, Alexander, 14

Foley, Jonathan, 47

food: ancestral, 130; cheap, 44, 45, 48, 53; colonization of, 40; consumption of, 64, 76, 98, 203; culturally relevant, 97, 219; distribution of, 42, 83–84, 184; foraged, 58; global, 60, 67, 100; growing, 58, 131; healthy, 42, 102, 104, 236; Mexican, 101, 102; preserved, 58; processed, 48; seasonal, 58; Tex-Mex, 101; Texas BBQ, 146
Food (Rollins and Bauknight), 24
food access, 8, 26, 41, 98, 181, 190, 212, 237; justice-focused, 42; systems of, 43
Food and Culture (writing course), 88, 193, 243
food apartheid, 41, 71
food choices, 26; ethical, 9, 131–32; systemic issues with, 105
food corporations, 40, 44, 45, 49, 53, 110, 135; marketing by, 50
food crisis, 28, 39, 110; writing about, 40–55
food culture, 102, 146, 147, 246; Indigenous, 37, 62, 80
food deserts, 41
food disparagement laws, 49
food forests, 89
Food, Inc. (documentary), 148
food insecurity, 19, 41, 42, 96, 129, 135, 181, 185, 188; disabilities and, 43
food justice, 8, 13, 26, 42, 49, 64, 66, 67, 129, 212, 213, 222; advocates, 61, 174, 246; Black, 174; fight for, 233; literature, 165; local, 68; organizations, 43; working for, 174
food labels, 50, 56
food libel laws, 49
food literacy, 10, 13, 23–24, 31, 32, 43, 66, 83, 92, 105, 111, 156, 159, 161, 181, 187, 213, 219; campaign for, 112, 113, 164, 167; climate change and, 128–32; collaborative, 106, 133; community, 8, 109, 138, 139, 166, 171, 180; control of, 53; critical, 24–25, 99; developing, 6, 106, 171, 188, 238; documents for, 193; goal of, 180; lack of, 49; projects, 5, 94, 107, 112, 189; relational abundance and, 236–43; rhetoric and, 25, 129; work in, 3, 4, 25, 27–28, 77, 195, 207, 220, 237

Food Marketing Institute, 86, 91
Food Matters (Bauer), 24
food miles, 87, 88, 94, 104
food production, 44, 46, 58, 76, 95; nonhazardous, 90; process of, 238; sustainable, 52, 53; system, 112; transparency in, 57; weather patterns and, 46
food rescue, 42, 158, 181
food security, 41, 51, 53, 95; community, 10; health issues and, 43
food shortages, 48; climate-related, 110; global, 60
food sovereignty, 26, 37, 67, 68–69, 211, 216; critical, 174; cultural, 213; Indigenous, 68, 208
food studies, 7, 11, 24, 39; advocacy with, 34–35; rhetorics and, 50
food systems, 13, 24, 64, 70, 72, 109, 129, 146; communicating about, 128; diversification of, 48; fear/anger and, 9; food justice and, 22, 174; global-scale, 62; multiple, 42; organic, 55, 155–56; problems with, 45, 59, 174–75; racist, 71; sustainable, 57, 134; understanding, 59; violence of, 243; visions for, 63. See also industrial food systems; local food systems
food waste, 5, 45, 97, 180, 181, 213
foodshed, 86, 132, 133, 190, 199, 209, 220
foodways, 9, 19; connecting, 131; cultural, 70; Indigenous, 37, 79, 98, 208, 217
"Forget Organic, Eat Local" (*Time Magazine*), 59
fossil fuels, 47, 109, 118
Foxglove Farm, 56
framework: antiracist/decolonial, 9; collective, 139; conceptual, 126; food justice, 207, 209, 236
Francis, Kerry, 185, 224, 226f, 233, 238, 241; distributed definition building, 237; grant for, 225; land and, 226, 227; learning from, 239–40; relational work and, 231, 232, 236; rhetorical agency and, 231; on systemic problems/solutions, 235
Francis, Tim, 185, 224, 233, 238, 241; distributed definition building and, 237; grant for, 225; land and, 226, 227, 229,

230, 232, 235; learning from, 239–40; relational work and, 231, 236; rhetorical agency and, 231
Freedom Dreams: Black Radical Imagination (Kelley), 136
Frontline Farming, 37, 67, 68, 130, 212, 213, 246
Fruit Rescue Harvest Festival, 181, 182*f*, 183*f*

García, Romeo, 20, 222
gardeners, physically disabled, 69
"Gardener's Dream: Dad Will Recover" (Weed), 201*f*; inspiration for, 200
gardens, 70; community, 43; family, 198; home, 158; market, 224; tending, 228, 229
Garth, Hanna, 174, 175
Gayeton, Douglas, 51, 65, 66
genetically modified (GM), 40, 78, 80
genetically modified organisms (GMOs), 40, 50, 52, 56, 78, 79, 80, 86, 100, 134, 135; labeling, 49; local and, 94; safety of, 136
genocide, 6, 63, 100, 141; cultural, 36, 148
Gerber, 51
gift economy, 67, 126–27
Gladwell, Malcolm, 160, 167
glyphosate, 52, 82
GM. *See* genetically modified
GMOs. *See* genetically modified organisms
Goldblatt, Eli, 16, 20, 32
Gordon, Constance, 163–64, 188; art contest and, 190, 191; field trips and, 191; food justice and, 164; teacher packets by, 189
Gorsevski, Ellen, 50
Gottlieb, Robert, 66, 86, 171
Gould, Jerome, 33
Gould, Marinda, 244
Gould family, 35, 36
Gould homestead, 244, 245*f*
Grabill, Jeff, 20
Grassroots Gardens, 69
Great Grief, 118
Great Law, 96
Green, Keisha, 18

greenhouses, 69, 192*f*
greenwashing, 50, 86
grief, 9, 32, 113, 121, 122; critical race theory of, 128; trauma and, 123, 124
Gries, Laurie, 190; on rhetoric, 25, 145, 162
Grieving While Black (Wade), 121
Growhaus, 68
Guattari, Félix, 140
Gubele, Rose, 220
Guinsatao, Terese, 241
Guthman, Julie, 66
Guttridge, Kena, 165
Guttridge, Mark, 165
Gwin, Pat, 48

Hahn, Laura K., 128
Hall, Madeline, 22, 240
handouts, 196*f*, 197*f*
Harlem Grown, 68
Harper, A. Breeze, 72
Harvest of All First Nations, 37, 80, 98, 211, 213
harvesting, 43, 185, 210
Hatherly Family, grant from, 213
Haudenosaunee Confederacy, 96
Hawhee, Debra, 110, 111
Hawisher, Gail E., 165
healing, 236; collective, 9*n*3; Indigenous, 242; process of, 243
health food market, 68
health issues, 26, 41, 52, 67, 94, 116; African American, 72; community, 71; food distribution and, 42; food security and, 43
heart disease, 41, 49
Hefferan, Kate: work of, 205*f*
Heinberg, Richard, 58
"Hen on Straw" (Kimball), 202*f*; inspiration for, 201–2
herbicides, 47, 52, 82, 87
Heritage Agriculture, 61
Hill, Cametria, 72
Hink, Gary, 14
Homestead Act (1862), 36
Honorable Harvest, 63, 246, 247
hooks, bell, 32, 207–8, 232
Horton, Miles, 135
House Agriculture Committee, 49–50

House Select Committee on Climate Crisis, 110
How Can Public Policy Create Opportunities and Barriers to the Development of a Local Food System? (Dancer), 152
H.R. 1599 (2015), passage of, 49–50
Hsu, V. Jo, 17, 107, 149
Hubrig, Ada, 18
humility, 16, 18, 128, 207, 241
Hummingbird Wholesalers, 69
Hundred Year Flood (1913), 35
hunger, 8, 57, 109, 110, 213; rates/poverty, 46
Hurricane Florence, 43

ideas: circulation of, 161–63; collective/collaborative, 137; place and, 22
identity, 130, 166; collective, 240; cultural, 150; food, 8; racial, 234
illiteracy, 26, 77; agricultural, 44; food, 31–32
imagination, 9, 116, 124, 232; radical, 21, 136, 137–38, 185
Indigenous peoples, 32, 35, 63, 78, 121, 130, 230, 234; acknowledging, 37; cultivation by, 36, 54; forced removal/genocide of, 36, 37; issues of, 211; ownership and, 6; violence against, 215
Indigenous Peoples Day, 211
Indigenous Peoples Projects, 212
"Indigenous Place-Thought and Agency Among Humans and Non-Humans" (Watts), 221
individualism, 16, 38, 63, 118, 128
industrial agriculture, 6, 50, 54, 55, 82; ecosystems and, 148; natural resources and, 47; USDA and, 60
industrial food systems, 41, 49, 59, 68, 73, 108, 127, 129, 172, 220; defining, 5–6, 39–50; dependence on, 40; discourses about, 39; emotional reactions to, 74; homogeneity of, 69; ingredients in, 105; problems with, 28, 45, 72, 148, 236; production of, 26, 148; racism/ableism/anti-Indigenous/practices in, 8; resources for, 52; responding to, 73; thinking about, 44

"Industry Report" (OTA), 55
inequalities, 62, 209; economic, 225; racial, 225; systems of, 72, 74
inequities: food, 174; income, 174; power, 67; racial, 213; structural, 43, 57, 65, 67
information: framing, 136; lack of, 50; suppressing, 124; traumatizing, 123
injustice, 9, 53, 122, 137, 184; antidote to, 5; food, 174; legacy of, 228; racial, 235; righting, 234
Inner Work of Racial Justice, The (Magee), 118
innovations, 63, 68–73, 74, 160
Institutional Review Board (IRB), 84n1
interconnections, 68, 147, 227, 233; radical, 126; root causes and, 149
International Agency for Research on Cancer, 82
intervention, rhetorical, 156–59
INVST program, 166
Irish Potato Famine (1845–1849), 48
Itchuaqiyaq, Cana Uluak, 18

Jackson, Rachel, 240
Jackson, Wes, 47, 60
Jefferson County Public Health, 213
Jones, Suzanne, 211
Joshi, Anupama, 67, 86, 171
Journal of Multimodal Rhetorics, The, 15
justice, 9, 21, 62, 63, 66, 109, 130, 184; climate, 116; community, 189; cultural, 98; decolonial, 222; defining, 172, 175, 207–14; disability, 18, 42, 43, 67; environmental, 152, 189; epistemic, 142; movement, 39; racial, 24, 43, 97, 98, 222, 233; work, 30, 71, 98, 142, 172. *See also* food justice; social justice

Kahneman, Daniel, 128
kapwa, 169, 241
Kawagley, Angayuqaq Oscar, 223, 224
Kelley, Robin D. G., 136
Kiehl, Jeffery, 116, 126, 135
Kimball, Shade: work of, 202f
Kimmerer, Robin Wall, 38, 65, 67, 68, 128, 218, 228; on animacy, 221–22, 223, 224, 227; on First Man / Second Man, 34; gift

economy and, 126–27, 184; on gratitude/abundance, 184; Honorable Harvest and, 63, 246; other-than-humans and, 223–24; on reciprocity, 227
King, Lisa, 36, 37, 220, 221
King, Matthew, 119, 126
Kingsolver, Barbara, 60
kinship, 6, 9, 19, 36, 140, 143; ecological, 25; reciprocal relation of, 63
"know your farmer" campaigns, 58, 59, 60
"know your food" campaigns, 59, 60, 190
knowledge(s): abundance of, 216; building, 14, 64, 99; co-creation of, 10, 247; community-based, 10; cultural, 130, 206, 210; ecological approach to, 219; hierarchies of, 224; Indigenous, 9, 140, 141, 147; multiple, 180; other-than-human, 220, 223; rhetorical, 219; sharing, 154, 207–8; teaching, 207–8
Koch, Megan, 86
Kunce, Catherine, 14
Kynard, Carmen, 124–25, 136, 139, 244

labor: abusive systems of, 53; cost of, 155; intensive, 156; land and, 235
LaDuke, Winona, 54, 71, 139, 141, 148
Lakoff, George, 114–15, 116
Lamar, Kendrick, 71
land: agricultural, 155; bodies and, 223; claims, 6, 35, 36, 40, 78; co-writing by, 215; colonial claiming of, 6; community and, 232; connecting to, 225; control of, 71; labor and, 235; living with, 227, 229; partnership with, 226; as pedagogy, 241; preservation of, 153; relationship with, 72, 98–99, 229, 230; rights, 70; tending of, 225, 235; theft, 26; treatment of, 221; white seizure of, 148; white-European-colonialist delusion of, 234
Land Institute, 47
land use, 157, 212; changing, 165; local food and, 98
Land Use Code, 149, 165
language, 8, 24, 35, 57, 68, 89, 221; co-creation of, 223; colonization of, 51; native, 230; writing and, 73

Latinx Writing and Rhetoric Studies, 15
Latour, Bruno, 140
Lertzman, Renee, 117, 118, 119, 123
Lexicon, The, 65, 66
liberation, 34, 68, 213
literacy, 135, 142, 151, 156, 170, 210; across communities, 14; collaboration on, 133; consumer, 158; critical ecology of, 166; cultural ecology of, 165; land-based, 211, 223; local food, 32, 157, 190, 195; projects, 154–55; public, 157, 217; violence of, 53; vision for, 32; women's, 18; work, 180, 184–85, 187
Literacy in American Lives (Brandt), 53
Literacy in Composition Studies, 15
Living Seed Library, 69, 70
Lobb, Richard, 46
Loberg, Lindsey, 41, 45
local: defining, 85, 87, 100, 105; GMOs and, 94; organic and, 58; power of, 190
local food, 4, 62, 65, 69, 73, 103, 108, 128, 158, 166; abundance and, 185; access to, 96, 176; affording, 157, 171; awareness of, 28, 156, 177; benefits of, 131; complexities of, 91, 133; concept of, 60–61, 90, 120, 162, 167; consumption of, 28, 38, 59, 60, 77, 105, 132, 133, 192, 209; defining, 3, 5, 10–16, 19, 21, 23, 25, 28, 32, 70, 76, 83–85, 87, 88, 92–93, 94, 96–99, 100, 104, 122, 130, 133, 145, 147, 159, 163, 169, 170, 171, 175, 176, 178, 180, 184, 187, 208; discussions of, 37, 96–97, 190, 207, 208; distribution of, 89, 178; ecology of, 10, 83, 101, 106, 132, 136, 177, 179, 181, 191; justice-focused, 213, 214; literacy about, 32, 157, 190, 195; production of, 26, 28, 38, 77, 81, 89, 132, 133, 147, 157, 171, 174, 178, 209; promoting, 99, 100, 133; purchasing, 85, 86, 91, 95, 99, 104, 105; restaurants and, 84, 85; studying, 38, 180; teaching about, 143; term, 4, 30; words/art and, 191; working for, 19, 170, 235; writing about, 101–6, 240
Local Food Art Competition, 190
"Local Food Movement 15 Years In, The" (Nestle), 60

local food movement(s), 21, 27, 29, 39, 66, 73, 98, 99, 113, 131–32, 167; challenges for, 76; defining, 58–63; pedagogy about, 75; success for, 129–30
Local Food Summit, 132
"Local Food and Sustainable Agriculture," 80–81
local food systems, 29, 64, 146, 149, 187, 209; building, 27, 77, 165, 218–19, 222; factors favoring, 155; just, 223
Local Food Think Tank, 100
Local Foods Wheel, 59
local trap, 62; salsa and, 99–106
locavore, 59, 63, 166
Longaker, Kenny, 61
loss, 19, 36, 121–22; experiencing, 123; restoring, 128
Love Our Community's Agricultural Life (LOCAL), 91

Macdonald, Christine, 14
Magee, Rhonda V., 118, 125, 126
Maiser, Jennifer, 59
Making Local Food Work, 132
Market Bucks, 91
Marshall, George, 114, 115, 118, 124
Martin, Trayvon, 234
Martinez, Tyler, 43
MASA Seed Foundation, 70
MASA Seed Project, 186
Massachusetts Center for Native American Awareness, 150
Mastrangelo, Lisa, 84
Mathieu, Paula, 16, 17, 32, 125
Matrix, The (film), 44, 51
Mayans, 242
McDonalds, 51, 62, 201
Meadows, Donella, 147, 148–49, 158
meat: cheap, 48; factory-farmed, 48; feedlot, 49; grass-fed, 49; local/organic, 85
meatpacking plants, 53
meatwashing, 50
media: discourse analysis of, 11; source documents for, 91; thematic clusters of, 90t. *See also* social media
Medina, Johnitta, 72

Menakem, Resmaa, 120, 121, 126, 242
Mentimeter, 213, 214
methodology, 7–8, 30, 75, 150, 161, 175; decolonizing, 221; ecological community writing, 19; relational, 11
Middendorf, Gerad, 40
Miele, Mara, 86
"Miles Matter. Eat Local" (Osofsky), 203*f*; inspiration for, 202–4
Millennial Ecosystem Assessment, 47
Minaj, Nicki, 71
Mitchell, Tania, 173–74, 209
Mohawk, 48
Momma Jah, 72, 213
Monberg, Terese, 32, 169
monocultures, 6, 48, 54, 68, 100, 148
Monsanto, 51, 54
Montinari, Massimo, 67
Morpheus, 44, 51
Moss, Beverly, 18, 32
Mother Jones, 49, 64
Mukavetz, Andrea Riley, 19, 141, 220, 234, 240
My Grandmother's Hands (Menakem), 120
Myers, Seth, 14

Nagel-Brice, Isaac, 206*f*
Nall, Madeleine, 23, 227, 228
National Center for Atmospheric Research, 116
National Chicken Council, 46
National Climate Assessment, 46
National Commission on Service-Learning, 173
National Organics Standard Board, 56
National Restaurant Association, 86
National Young Farmers Coalition, 154
Native Americans. *See* Indigenous peoples
Native Nations, 98, 212, 217
Native Seed Search, 70
Natural Discourse (Dobrin and Weisser), 143
natural resources: industrial agriculture and, 47; limited, 51
nature, 232; relationship with, 233, 235
neonicotinoids, 134

Nestle, Marylin, 60
Neustras Raices, 68–69
new materialism, 6n2, 7, 140, 141, 142, 143
New York Times, 49, 64
Newton, Peter (Pete), 152, 153, 156, 157, 163
No Cost Groceries, 181
Nord, Mark, 43
"Nose to Tail" (Nagel-Brice), 206f; inspiration for, 204–6
"Nothing About Us Without Us" (Hubrig), 18
Nowacek, David, 54, 55
Nowacek, Rebecca, 54, 55
nutrients, 47, 82; calories and, 41; cancer-fighting/health-promoting, 87

Office of Outreach and Engagement, 13, 152, 172, 225
Oglala Lakota, 130
Omnivore's Dilemma, The (Pollan), 64
Open Space Cropland Policy, 153
oppression, 19, 116; dismantling, 150; systemic/institutional, 209
Ore, Ersula, 234, 245
organic, 148; certified, 56, 87; defining, 54–58; local and, 58; term, 56
organic agriculture, 51, 81, 97, 98, 153, 155; defining, 54; environmental benefits of, 55; importance of, 58; integrity of, 56
organic food, 44, 58, 156; accessibility of, 57; consumption of, 54; local, 85; production of, 26
organic movement, 39, 54, 56, 129
Organic Trade Association (OTA), 55
Osofsky, Ellia: work of, 203f
other-than-humans, 30, 140, 143, 147, 177, 220, 222, 223–24, 227, 237, 241, 247; agency of, 219; co-writing by, 215; collaboration with, 231; including, 229; as partners, 240; partnership with, 221; personhood and, 224; potential of, 243
Owens, Derek, 32, 107

Pacheco-Borden, Carmen, 101, 102, 103–4
Parkin, Micah, 180, 181, 182f, 186
Parks, Stephen, 20, 32

partnerships, 13, 19, 163, 221, 232, 236; building, 219, 226, 231; co-creative, 228; community, 219; ecology of, 216; efficacy of, 20; reciprocity and, 220, 241–42; trans-community, 219; writing, 228
Pecoraro, Richard, 186f
pedagogy, 22; engaged, 12, 32
Peña, Devon, 211, 220, 231
Penniman, Leah, 71; charitable system and, 41–42; food justice and, 41; Soul Fire and, 130
People's League for Action Now (PLAN-Boulder), 154
"Perceived Opportunities for, and Barriers to, the Development of Local Food Systems: A Case Study from Boulder County, Colorado" (House and Dancer), 152
personal work, 19; systemic work and, 243–47
personhood, 6n2, 221, 224
pesticides, 47, 52, 53, 54, 56, 57, 82, 87, 100
Pew Research, 72
Pezzullo, Phaedra, 152, 164
place, 146; history of, 169; ideas and, 22; work and, 34
plants, 56; co-writing by, 215
Plass, Tim, 132
Platt, Marinda May, 32–33, 34, 35–36, 38
Pleasure Activism (brown), 9n3
politics, 15, 63, 72; environmental, 144
Pollan, Michael, 39, 40, 64, 66, 68, 82; eating right and, 65; on feedlot system, 48–49; on food industry, 45–46; renaissance/remember and, 65
pollution, 229, 233; air, 53; water, 53, 111
Post Carbon Institute, 58
poverty, 8, 46, 96
Powell, Malea, 141
power, 35, 74, 128; collective, 24; community-based, 26; corporate, 26; definition-building, 5–8, 175, 241; dynamics, 5, 122; hierarchies of, 137; imbalances, 39, 247; organizational, 21; political, 42; transformative, 38
Prentice, Jessica, 59, 63
preservation, 181, 184; food, 100, 180,

224–25; increasing, 132; land, 80
Primm, Annelle B., 120
psychological studies, 27, 109, 113, 117
psychology, 123, 127; behavior and, 112; clinical, 116; trauma, 120, 121. *See also* ecopsychology; social psychology
public writing, 17, 25, 151; commitment to, 18; events, 28; projects, 161, 164, 187
Purcell, Mark, 62, 100

race, 96, 110, 120, 122, 137, 190, 208
racism, 8, 38, 41, 120, 174–75, 207, 234–35; agriculture and, 233; environmental, 120; reinforcing, 243; systemic, 52, 57, 59
Rarámuri, native language of, 230
Rathvon, Sally, 32
"Raw" (Hefferan), 205*f*; inspiration for, 204
Raymond, Spencer, 33
Ready to Work program, 172
reciprocity, 16, 17, 19, 21, 219, 247; cultivating, 20, 224–36; Indigenous scholarship on, 29; intimacy/partnership and, 241–42; long arc of, 169; triangle of, 237
reclamation: community, 69; cultural, 69, 130; in definition building, 68–73
Reese, Ashanté M., 174, 175
Reflections, 15
regenerative considerations, 81, 165, 172, 208, 216, 229
Reiley, Laura, 86
relational abundance, food literacy and, 236–43
relationality, 19, 143, 145, 231; concepts of, 141, 223
relationships, 9, 10, 19, 88, 147, 184, 206, 232, 240; authentic, 236; building, 93–94, 96, 229, 247; co-creation of, 223, 227; colonial, 228; ecology of, 20, 238; economic, 65; intersectional, 120; with land, 72, 98–99, 229, 230; reciprocal, 231, 237; social, 126; strains on, 113; systems of, 216; tight-knit, 87
"Remapping Settler Colonial Territories: Bringing Local Native Knowledge into the Classroom" (Anderson), 218

research, 4, 16, 19, 30, 37, 217, 243; action, 11, 76; archival, 14; cognitive, 128; community-based, 42; environmental, 128; ethnographic, 15; interdisciplinary, 13, 143; local, 156–59; psychological, 113; public, 11, 28, 161; trauma-informed, 27; writing, 11, 83–88
resilience, 10, 70, 97, 116, 175; community, 69; environmental, 188; local food, 60, 109
resources, 64; abundance of, 216; accumulation of, 221; sharing, 213; systems of, 216
Restaurant Industry Association, 86
restaurants, 104, 161, 171; farm-to-table, 85, 87; full-circle, 86; local food and, 84, 85, 110
Restore Colorado program, 81
Rewriting Partnerships (Shah), 20
Rhetoric and Writing Studies, 3–4, 11, 13, 16, 24, 25, 39, 42, 45, 64; community-engaged, 19; ecological community writing and, 12; environmental issues and, 107; Indigenous scholars and, 127; potential of, 247; practitioners of, 4, 6, 7
Rhetoric, Politics, and Culture, 15
Rhetoric Society of America, 24
rhetorical concepts, 63, 147, 171
rhetorical contagion, 143, 160, 177
rhetorical life, 25, 28, 221, 236; other-than-human, 223
rhetorical overwhelm, 112–20
rhetoric(s), 79, 89, 97, 108, 110, 161; adaptive, 135, 138; advocacy, 67; antiracist, 150; circulation of, 162–63; colonial, 78; crisis, 111; critical, 109; cultural, 17, 150; deficiency, 128, 129, 132–38, 196; ecological, 7, 141, 143–45; evolution of, 159; farmer support, 97; fear-based, 135, 171; food, 3, 24, 97, 109; food literacy and, 25; food studies and, 50; Indigenous, 150, 220; local food, 25, 27, 28, 75, 91, 140, 161, 209; logic-/values-based, 63–73; medicinal, 228, 242; new materialist, 7, 143–45; other-than-human, 231; scarcity, 111
Rice, Jenny, 112, 129, 135, 136

Richardson, Elaine, 18, 150
Riedel, Charlie, 50
Right Relationship Boulder, 37
Ríos, Gabriela, 220, 223, 224
Ritual Chocolate, 103
Rocky Mountain Seed Alliance, 70
Rodale Institute, 52
Rodale Press, 54
Rodgers, Ross, 69, 199f
Rosebud Sioux Native Nation, 69
Rosenberg, Lauren, 21
Rosiek, Jerry Lee, 141
Roundup, 52, 54
Royster, Jacqueline Jones, 136, 137, 139
Ruiz, Iris D., 221, 236; historical curanderisma and, 242; medicinal rhetorics and, 228, 242

Salatin, Joel, 46, 48
Salmón, Enrique, 41, 230
salmon fisheries, 87
salmonella, 44
salsa, 102; farmers market, 101; producing, 103
Sampson, Dede, 59
Sand Creek Massacre (1864), 33, 36, 37, 211
saturation, 175–81, 184–85, 187–91, 193–207
scarcity, 38, 111, 128, 243
Schell, Eileen, 44, 60, 67; on GMOs, 50; on local/regional food, 100
Schlagel, Paul, 134, 137
Schlosser, Eric, 100
Science, 47
Scoop, The (blog), subjects on, 178–79
Seas, Kristen, 158, 159, 160, 167
seed banks, 70, 195
seed companies, 44, 135, 220
Seed Room CSA, 68
Seed Savers Exchange, 70
seed saving, 58, 69, 70, 210, 225; workshops, 185, 187
seeds, 195, 199; co-writing by, 215; colonization of, 40; GMO, 56; heritage, 71
Seeds for the People, 70
service-learning, 7, 14, 16, 32; critical, 83, 173–74

"Serviceberry: An Economy of Abundance, The" (Kimmerer), 184
settler colonialism, 33, 34, 36, 71, 79, 109, 147, 148, 174, 208, 234–35, 237, 240, 243, 245; agriculture and, 80, 233, 235
Shah, Rachael W., 20, 223
Shamrock, 85
Shed: Boulder County Foodshed, The, 111–12, 132, 136, 143, 153, 161, 164, 165, 168, 176, 206, 208; art competition by, 190; board of, 133, 134, 141, 156–57, 158, 167, 171, 175, 188, 193, 211, 215; brochure by, 194f, 195f; common agenda and, 216; consumers and, 156; discussions with, 178; ecology of, 210; education campaign by, 166; events/workshops by, 188; goals of, 171–72, 177; local food and, 152, 166, 187; social media and, 178–79, 193; website of, 178, 190, 193
Sherman, Sean, 130, 131, 220
Shiva, Vandana, 39, 54, 69
Simpson, Leanne Betasamosake, 241
Simpson, Natasha, 43
Sioux Chef's Indigenous Kitchen, The (Sherman), 130
Sistah Vegan, 72
skills workshops, as literacy work, 180, 184–85
slavery, 5, 6, 26, 53, 233, 234
Sligh, Michael, 56
Slow Food Boulder County, 102
Smith, Linda Tuhiwai, 221
Smith, Mistinguette, 230
Smithfield, Inc., 201
SNAP (Supplemental Nutrition Assistance Program), 91, 97
social issues, 4, 5, 6, 12, 25, 76, 95, 108, 114, 122, 123, 166, 179, 210; conventional/linear thinking and, 149
social justice, 67, 109, 116, 121, 125, 189, 211
social media, 193, 206, 207; literacy campaign and, 166; using, 178–79
social psychology, 20, 117, 118–19
soil depletion, 48, 111, 148
soil health, 56, 138, 149, 155, 157
Soul Fire Farm, 71, 130

Southern Girl's Guide to Plant-Based Eating, A (Hill), 72
sovereignty, 9, 68, 212; community, 18; cultural, 8, 24, 37, 72, 96, 116, 213; movements, 130; seed, 69, 70; Tribal Nation, 212. *See also* food sovereignty
soybeans, 48; GM, 49, 52
spaces: academic, 15, 18; community, 18; environmental, 89; public, 187; urban growing, 58
Spark, 15
Spears, Patrick, 139, 141
Spirit of the Sun, 37
Square Roots, 68
Staff Land Acknowledgment, 78, 211
stewardship: ethical, 224; generational/heritage pride of, 79
Still Life with Rhetoric (Gries), 145
Stoknes, Per Espen, 116, 118, 218
Stroh, David Peter, 149, 158
Stuckey, Elspeth, 53
sugar beets, GM, 79, 80, 81, 154
sustainability, 51, 52, 53, 54, 57, 62, 67, 82–83, 86, 172, 210; defining, 78–79; economic, 81; encouraging, 81; environmental, 81; importance of, 82, 84; land and, 80; long-term, 81; sourcing, 85
Sustainable Food Center, 69
Sustainable Food Systems, 157
Sweet Potato Soul, 72
Sysco, 84, 85, 87
systemic issues, 107, 109, 175, 210, 235
systemic work, personal work and, 243–47
systems, 116, 165; change, 213; global, 211; interlocking, 214; nonlinear, 158; racist, 174–75; studying, 151; theory, 7, 28, 143, 162, 176; writing, 147–50
Systems Thinking for Social Change (Stroh), 149
Syverson, Margaret, 144

TallBear, Kim, 141
Taylor, Brianna, 233
Tekobbe, Cindy, 220, 234, 240
10 Million Black Vegan Women, 72
terroir, 145–46, 147

Thinking in Systems (Meadows), 147
Thompson, John R., 87, 166, 167
Till, Emmett, 234
Time Magazine, 59
Tipping Point, The (Gladwell), 160
Todd, Zoe, 141
Tony's Market, 86
trauma, 111, 113, 116, 134; ancestral, 121, 122; critical race theory of, 128; genetic expression and, 121; grief and, 123, 124; intergenerational, 120, 121; legacy of, 235; studies on, 109
trauma studies, 139, 176; lessons from, 112–20
Trauth, Erin, 51
Tribal Nation Ethnographic-Education Report, 212
Tribal Nations, 98, 212

UN Climate Change Annual Report, 46, 47
UN International Panel on Climate Change, 46
Union of Concerned Scientists, 48
University of Colorado Boulder, 31, 32, 83, 151, 152, 157, 164, 166, 173, 212, 213; food-focused writing courses and, 132; writing program at, 12
University of Denver, 12, 212, 243, 246; writing program at, 83, 213
University Writing Program (Denver), 83, 213
Urban Roots, 68
US Foods, 85
USDA, 40, 56; food production and, 58; industrial agriculture and, 60; organic food market and, 55
USDA Economic Research Service, 42
USDA Organic, 56
Utes, 35
"Utilizing Art for Social Change" (Constance), 189

Valmont School, 33, 35
Van Sustern, Lise, 113
Van Wing, Sage, 59
Vegan Voices of Color, 72

veganism/vegetarianism, 72
veggie libel laws, 49
viability, 156; economic, 135; farm, 27, 97, 152, 158
Vibrant Earth Seeds, 70
Village Market, 68
violence, 53, 233, 242, 245; ancestral, 122; colonial, 40; food-related, 53, 243; against Indigenous peoples, 215; legacies of, 19, 34, 37, 52, 234; perpetuation of, 235; racial, 234; resisting, 33
virality, theories of, 143, 159–61

Wade, Breesha, 125; on imagination, 137–38; on loss, 121–22
Walmart, 61, 62, 65
"Want to Get Involved . . ." (flier), 196
Washington, Karen, 41
water, 47, 216; availability of, 138, 149; colonization of, 40, 40n1; control of, 71, 110
water rights laws, co-writing by, 215
Watts, Vanessa, 221
weather patterns, food production and, 46
Weed, Jules: work of, 201f
Weisser, Christian, 144, 145
wellness, critical race theory of, 128
Wenger, Paula, 187
"We've Forgotten How to Listen to Plants" (Kimmerer), 224
What We Think About When We Try Not to Think about Global Warming (Stoknes), 218
"What Will We Eat as the Oil Runs Out?" (Heinberg), 58
Whatley, Booker T., 54
"When I Eat Raspberries I Feel Like a Fairy Princess" (Ellis), 200f; inspiration for, 199
White, Rowen, 48
white supremacy, 24, 35, 41, 111, 118, 120, 122, 216, 243

whiteness, 43, 66, 172, 242
"Why Interspecies Thinking Needs Indigenous Standpoints" (TallBear), 141
WIC (Women, Infants, and Children), 69, 91, 97
wild places, 229, 230, 233
WISE. *See* Writing Initiative for Service and Engagement
Word Cloud, 214f
World Health Organization, 52
Wright, Wynne, 40
writing: business, 12; collaborative, 29, 127–28, 175, 219; critical, 83; culture and, 145; as distributed process, 162; dynamic ecology of, 215; Indigenous, 142, 233; language and, 73; as "liquid" system, 162; multimodal, 179–80; nature, 144; reflective, 207; science, 12; systems, 147–50; theories of, 24–25, 140. *See also* community writing
Writing Across the Curriculum, 30
writing courses, 14, 28, 30, 140, 187; community-engaged, 12; theory/research and, 13
Writing Democracy (Carter and Mutnick), 15
"Writing Ecologies, Rhetorical Epidemics" (Seas), 159
"Writing for Food Justice: Harvesting Hope in Abundance," 212, 213
Writing Initiative for Service and Engagement (WISE), 12, 13, 173
Writing and Rhetoric Studies, 129, 175

Yards to Gardens, 166
Young, Anna "Amy," 146

Zen Buddhism, 122
Ziibimijwang Farm, 68

About the Author and Contributors

VERONICA HOUSE is a professor in the University Writing Program at the University of Denver. She has published several articles on local food movements and community engaged work in higher education. She is the founding director of the Conference on Community Writing and founding executive director of the Coalition for Community Writing. Veronica has served as coeditor of the *Community Literacy Journal* since 2016. She is recipient of Campus Compact of the Mountain West's Engaged Scholar Award; University of Colorado's Women Who Make a Difference Award; and numerous teaching awards.

* * *

KERRY FRANCIS received her BS in biology from University of Puget Sound and her MA in counseling psychology from Naropa University. Kerry and her husband cofounded Dharma's Garden, a nonprofit educational farm within the City of Boulder, with a mission to demonstrate, educate, and inspire others to take on the noble work of tending the earth. On their biodynamic farm, Kerry is the director of education, creating and teaching land-based, nature-inspired curriculum for classes and workshops for young children, teens, and adults. She is also a certified Waldorf preschool teacher, as well as a professional editor and certified birth doula.

TIM FRANCIS is the executive director and co-founder of Dharma's Garden, a nonprofit educational farm within the City of Boulder. A Boulder native, Tim is a lifelong gardener, landscaper, and nature lover. Before founding Dharma's Garden, Tim studied biodynamic farming under Jim Barausky in Boulder and then honed his skills internationally as the garden manager at Koanga, a permaculture institute in New Zealand. Tim also holds credentials as a certified arborist through the International Society of Arboriculture.

CONSTANCE GORDON is an assistant professor in the Department of Communication Studies at San Francisco State University. She is a core faculty member in SF State's transdisciplinary Climate Change Certificate Program and Climate Justice Leaders Initiative. Gordon's research addresses the rhetoric and critical organizing practices of food and environmental justice movements, such as how organizers challenge multiple forms of dispossession, articulate intersectional political critique, and facilitate mutual aid even amid repression. Her solo and collaborative writing can be found in *Environmental Communication, Cultural Studies, Quarterly Journal of Speech, Journal of Applied Communication Research, Frontiers in Communication*, and numerous chapters in edited volumes. She is one of the co-authors of the textbook *Rhetorical Histories of Social Movements in the U.S.* (Cognella, 2025).

KELLY ZEPELIN is a cultural anthropologist whose research explores the intersections of wild food foraging, ethics, and ecological and human health. She completed her PhD at CU Boulder, where her work investigated contemporary wild food traditions in the American West, with a focus on the relationships between foragers, plants, and landscapes. Her research integrates perspectives from soil microbiology, decolonization, and the cultural significance of food. A recipient of numerous awards, including the David M. Schneider Award from the American Anthropological Association, Kelly's work highlights the cultural and ethical dimensions of wild food practices. She is working to become a nature-based counselor, combining a deep connection to the natural world with her commitment to supporting human-ecological well-being.

About the Cover Artist

PHOEBE DRAPER is a botanical illustrator and farmer, hailing from the foothills of the Blue Ridge Mountains of Appalachia. After studying architecture and landscape architecture for years, their work has transitioned to the physical realm, with their hands and pen as the primary tools they use to connect with Earth. Botanical illustration allows Phoebe to build relationships with plants and ecosystems by slowing down and getting curious. Their work explores concepts of bioregionalism, ecosystem succession, and soil habitats, often seeking to question and traverse the divide we draw between natural ecosystems and vegetable farming. A traveler at heart, Phoebe has spent time in varying ecosystems and always finds a similar theme at the core of their journey: building connection through reciprocity with the natural world. This most often is accomplished with a pen in hand.

www.ingramcontent.com/pod-product-compliance
Lightning Source LLC
Chambersburg PA
CBHW050223100526
44585CB00017BA/1800